MATH FOR LIFE AND FOOD SERVICE

Lynn Gudmundsen

Prentice
Hall

Upper Saddle River, NJ 07458

Library of Congress Cataloging-in-Publication Data

Gudmundsen, Lynn.
   Math for life and food service / Lynn Gudmundsen.
      p. cm.
   ISBN 0-13-031937-6
      1. Mathematics.   2. Business mathematics.   3. Cookery--Mathematics.   I. Title.
   QA39.3.G83 2002
   510--dc21

                                                            2001036435

Editor-In-Chief: *Steve Helba*
Executive Editor: *Vernon R. Anthony*
Production Management: *Patty Donovan, Pine Tree Composition*
Production Liaison: *Barbara Marttine Cappuccio*
Director of Production and Manufacturing: *Bruce Johnson*
Managing Editor: *Mary Carnis*
Manufacturing Manager: *Cathleen Petersen*
Creative Director: *Cheryl Asterman*
Cover Design Coordinator: *Miguel Ortiz*
Cover Designer: *Amy Rosen*
Cover Image: *Eyewire / PhotoDisc Photography*
Marketing Manager: *Ryan DeGrote*
Editorial Assistant: *Anne Brunner*
Composition: *Pine Tree Composition*
Printing and Binding: *Banta Book Group*

Photo Credits: © Darling Kindersley, Ltd. (2) David Murray and Jules Selmes. (3) David Ward.
(4) Jerry Young. (5) Susanna Price. (6) Jan O'Leary. (7) David Murray. (8) Philip Wilkins.

Pearson Education Ltd.
Pearson Education Australia Pty, Limited
Pearson Education Singapore, Pte. Ltd.
Pearson Education North Asia Ltd.
Pearson Education Canada, Ltd.
Pearson Educación de Mexico, S.A. de C.V.
Pearson Education—Japan
Pearson Education Malaysia, Ptd. Ltd.
Pearson Education, Upper Saddle River, New Jersey

Prentice
Hall

10 9 8 7 6 5 4 3 2 1
ISBN 0-13-031937-6

Contents

Appendices

Preface

A NOTE TO THE STUDENT

This text assumes that you, the student, have had a basic course in arithmetic skills including addition, subtraction, multiplication, and division of whole numbers, decimals, fractions, and percents. The coverage of these topics is not intended to be complete. For a complete and thorough understanding of decimals, fractions, and percents, you should consult a text dedicated to teaching a complete knowledge of these topics.

In addition, this textbook does not teach every mathematical skill needed in the food service industry. There are numerous computational skills involved in running a food service business including business taxes, employee payroll, payroll taxes, mortgages, interest rates, amortization, and depreciation, just to name a very few. What this textbook does offer you is a great deal of good examples of mathematical skills you can use in your day-to-day life as well as in your pursuit of a career in the food service industry.

Mathematics is considered a discipline and thus needs to be practiced daily. Organization is one key to success with mathematics. By keeping your thoughts and work orderly, your thinking will become more clear and precise. The quality of your work will improve greatly.

Use the examples in this text as your teacher. Read an example, see how it is worked out, then cover it up with a piece of paper. Try to recreate the solution. If you get stuck, peek . . . then cover it up and try again. Keep up this process until you can successfully recreate the solution. I guarantee that if you don't just give up and go on to the next example, this technique will program your computer brain. You will have success!

THANK YOU'S

I would like to acknowledge the help of my colleagues here at Maui Community College. Kathy Acks spent many hours of her time and gave willingly her expertise in editing this manuscript. To Bobby Santos, Karen Tanaka, David Thielk, and the MCC "50H" class that first used this text, thanks for the edits. Your comments and support were invaluable.

I was also fortunate to have the help of a renowned local chef, Peter Merriman. He is the head chef and owner of the Hula Grill located in the Whalers Village in Kaanapali.

Yet another wonderful resource came to me via the internet, when I found the National Baking Center website. Didier Rosada, a baking instructor at the NBC, kindly answered all of my questions concerning the need for and use of baker's formulas (Unit 4).

Sharon and Dick Dobner, owners of the Sister Bay Bowl, were an absolutely wonderful resource. They spent many hours sharing their thirty years of restaurant and business experience with me. They have a wonderfully clean kitchen and great food at the Sister Bay Bowl. Thanks so much for allowing me to take unannounced pictures of your kitchen! To Susan Crossiant, owner of Sweetie Pies, your personal Excel business forms were wonderful.

Last, but certainly not least,—to the contrary—the most help came from my family. First, my sister Gail. Not only did she make it possible for me to skip teaching summer school to write this book, but she bought the laptop computer on which this very manuscript was created. Secondly, my mom Virginia, who spoiled me rotten while I was working 17 hours per day. Without her, I would not have eaten nor have had any clean clothes. Worst of all, my son would have fired me as his mom! THANK YOU.

A Review of Fractions, Decimals, and Algebra

This unit will not attempt to teach you how to perform mathematical operations with fractions, decimals, or percents. Neither will it attempt to teach you the concepts of algebra. This unit is for REVIEW. The word review implies that you have learned the material in the past, but you need a bit of refreshing. Please refer to your old math books if you need more detailed instructions for any of these reviewed procedures.

I suggest that you purchase a scientific calculator that has a fraction key a ⬚ and learn to use it. Accuracy is extremely important in any business. The knowledge of how to perform arithmetic calculations, combined with the correct use of a calculator, should greatly improve your accuracy.

CHAPTER 1

Fraction Review

$$\frac{numerator}{denominator}$$

Basic fraction addition/subtraction involves either fractions with common denominators, or fractions without common denominators. If the fractions you need to add/subtract have common denominators, you simply add the numerators and put your answer over the common denominator, reduce, and you are **pau** (Hawaiian for finished!).

If however, the denominators of the fractions you are attempting to add/subtract have different denominators, you must rename the fractions to insure that they do indeed have common denominators before you begin the addition process.

A. $\quad \dfrac{3}{5} + \dfrac{4}{5} = \dfrac{7}{5} = 1\dfrac{2}{5}$ *These fractions have a common denominator.*

B. $\quad \dfrac{2}{3} + \dfrac{1}{2} = \dfrac{2}{3} x \dfrac{2}{2} + \dfrac{1}{2} x \dfrac{3}{3} = \dfrac{4}{6} + \dfrac{3}{6} = \dfrac{7}{6} = 1\dfrac{1}{6}$

These fractions did not begin with a common denominator, so I used a simple trick to get a common denominator. See if you can spot the trick!

C. $\quad \dfrac{2}{3} - \dfrac{1}{4} = \dfrac{2}{3} x \dfrac{4}{4} - \dfrac{1}{4} x \dfrac{3}{3} = \dfrac{8}{12} - \dfrac{3}{12} = \dfrac{5}{12}$

Remember to perform the multiplication before the subtraction.

D. $5\dfrac{2}{3} - 2\dfrac{3}{5} = \qquad 5\dfrac{2}{3} = 5\dfrac{10}{15}$ *The fractions have been renamed as*

$\qquad\qquad\qquad\qquad -2\dfrac{3}{5} = 2\dfrac{9}{15}$ *a common denominator is necessary.*

$$\dfrac{}{3\dfrac{1}{15}}$$

E. $3\dfrac{1}{2} - 2\dfrac{2}{3} = \qquad 3\dfrac{1}{2} = 3\dfrac{3}{6} = 2\dfrac{9}{6}$ *Borrow* $1 = \dfrac{6}{6}$ *from the 3.*

$\qquad\qquad\qquad\qquad -2\dfrac{2}{3} = 2\dfrac{4}{6} = 2\dfrac{4}{6}$

$$\dfrac{}{\dfrac{5}{6}}$$

When multiplying fractions, first multiply the numerators and then the denominators. If you can cross cancel first, do so, then continue to multiply. This is a task that requires neatness and legibility when canceling, to avoid errors. An improper fraction (one with a numerator greater than the denominator) needs to be converted to a mixed number.

F. $\quad \dfrac{1}{2} \times \dfrac{1}{3} = \dfrac{1}{6}$

G. $\quad 1\dfrac{2}{3} \times 15 = \dfrac{5}{3} \times \dfrac{15}{1} = \dfrac{5}{\cancel{3}_{1}} \times \dfrac{\cancel{15}^{5}}{1} = 25$ *Cross cancel the 15 and the 3.*

H. $\quad \dfrac{1}{4} \times \dfrac{2}{5} = \dfrac{1}{\cancel{4}_{2}} \times \dfrac{\cancel{2}}{5} = \dfrac{1}{10}$ *The 2 and the 4 have been cross canceled.*

I. $\quad \dfrac{3}{5} \times \dfrac{10}{21} = \dfrac{\cancel{3}^{1}}{\cancel{5}_{1}} \times \dfrac{\cancel{10}^{2}}{\cancel{21}_{7}} = \dfrac{2}{7}$ *Both fractions have been cross canceled.*

J. $\quad \dfrac{11}{16} \times \dfrac{4}{5} = \dfrac{11}{\cancel{16}_{4}} \times \dfrac{\cancel{4}^{1}}{5} = \dfrac{11}{20}$ *The 4 and the 16 have been reduced (cross canceled).*

Division is the inverse of multiplication, and to divide fractions you must multiply! Remember the word **reciprocal**? Simply put, it means to turn a fraction upside down. Mathematically speaking, two numbers whose product is 1 are called reciprocals.

K. $\dfrac{4}{3}$ is the reciprocal of $\dfrac{3}{4}$ because $\dfrac{4}{3} \times \dfrac{3}{4} = 1$

L. $\dfrac{1}{2}$ is the reciprocal of 2 because $\dfrac{1}{2} \times \dfrac{2}{1} = 1$

To divide fractions, multiply by the reciprocal of the divisor.

M. $\dfrac{3}{4} \div \dfrac{2}{5} = \dfrac{3}{4} \times \dfrac{5}{2} = \dfrac{15}{8} = 1\dfrac{7}{8}$

Remember to multiply by the reciprocal of the fraction on the right (the divisor), <u>not</u> the fraction on the left (the dividend).

N. $\dfrac{7}{8} \div 6 = \dfrac{7}{8} \div \dfrac{6}{1} = \dfrac{7}{8} \times \dfrac{1}{6} = \dfrac{7}{48}$

O. $24 \div \dfrac{3}{8} = \dfrac{24}{1} \div \dfrac{3}{8} = \dfrac{\overset{8}{24}}{1} \times \dfrac{8}{\cancel{3}} = \dfrac{8 \times 8}{1} = 64$ *Cross cancel the 24 and 3.*

P. $\left(\dfrac{1}{4} + \dfrac{1}{3}\right) \div \dfrac{3}{16} = \left(\dfrac{1}{4} \times \dfrac{3}{3} + \dfrac{1}{3} \times \dfrac{4}{4}\right) \div \dfrac{3}{16} = \left(\dfrac{3}{12} + \dfrac{4}{12}\right) \div \dfrac{3}{16} = \dfrac{7}{12} \div \dfrac{3}{16}$

$\Rightarrow \dfrac{7}{12} \div \dfrac{3}{16} = \dfrac{7}{12} \times \dfrac{16}{3} = \dfrac{7}{3} \times \dfrac{4}{3} = \dfrac{28}{9} = 3\dfrac{1}{9}$

CHAPTER 2

Decimal Review

Keep the decimal point lined up for addition, subtraction, and division. This will ensure accuracy.

A. 3.251 + 0.0045 = 3.2555   *Remember to line up the decimal points, and use 0 as a place holder when necessary.*
    Write 3.251 as 3.2510 before you add 0.0045. This will insure accuracy.

$$
\begin{array}{r}
3.251\mathbf{0} \\
+0.0045 \\
\hline
3.2555
\end{array}
$$

B. 146.503 − 4.6 = 141.903   *Using zero place holders, rewrite 4.6 as 4.600 before you begin.*

$$
\begin{array}{r}
146.503 \\
-\quad 4.6\mathbf{00} \\
\hline
141.903
\end{array}
$$

C. 5.8 − 3.096 = 2.704   *Rewrite 5.8 as 5.800 before you begin. Then borrow from the 8 as usual.*

$$
\begin{array}{r}
5.\,8^{\mathbf{7}}\!\!\!\!\;\cancel{0}^{\mathbf{9}}\!\!\!\!\;{}^{1}0 \\
-3.0\,9\,6 \\
\hline
2.7\,0\,4
\end{array}
$$

D. 0.28 × 7.409   ⟹

$$
\begin{array}{r}
7.409 \\
\times\quad 0.28 \\
\hline
59272 \\
14818\text{«} \\
0000\text{««} \\
\hline
2.07452
\end{array}
$$

First multiply as you would for whole numbers.

Add the number of places to the right of the decimal points (3 + 2 = 5).

Place the decimal point 5 places in from the right.

E. $5.2\overline{)44.252}$ $\underset{\textit{divisor)}\overline{\textit{dividend}}}{\overset{\textit{quotient}}{}}$

Before you begin to divide, move the decimal point in the divisor (5.2) one place to the right to make it 52. Then, do the same to the dividend (44.252) to make it 442.52.

Next, bring the decimal point straight up into the quotient for your final answer.

$$
5.2\overline{)44.2.52} \quad \Rightarrow \quad
\begin{array}{r}
8.51 \\
52\overline{)442.52} \\
\underline{416} \\
265 \\
\underline{260} \\
52 \\
\underline{52} \\
0
\end{array}
$$

Keep in mind that you always move the decimal point in the divisor to the right enough times to make it a whole number. You must then move the decimal point in the dividend to the right an equal number of times before you divide.

Mathematically, you are multiplying both the divisor and dividend by an equal number of 10s to make the divisor a whole number and not change the value *of the ratio.*

F. $2.5\overline{)4}$ *Move the decimal point in the 2.5 over once to make it 25 (a whole number), then do the same to the 4, making it 40.0.*

$$
\begin{array}{r}
1.6 \\
25\overline{)40.0} \\
\underline{25} \\
150 \\
\underline{150} \\
0
\end{array}
$$
   Check: 2.5 x 1.6 = 4.0 = 4 ✓

For this division problem it was necessary to add an extra zero to the dividend, to get the answer to "stop" or "come out even." It has **no remainder**. Sometimes when dividing decimals, you have to add many zeros for the answer (quotient) to finally come out even.

The decimal representation of a rational number either ends or repeats. When the quotient repeats a pattern of numbers, it is called (*no imagination necessary*) a **repeating decimal**. The decimal representation of an irrational number such as π neither ends nor repeats.

G. $\dfrac{5}{9} \Rightarrow 9\overline{)5} = 0.\overline{5}$

$$\begin{array}{r} 0.555 \\ 9\overline{)5.000} \\ 45 \\ \hline 50 \\ 45 \\ \hline 50 \\ 45 \\ \hline \end{array}$$

Note that the 50 will reoccur continuously. To indicate repetition, put a line above the repeating numeral. Written: 0.$\overline{5}$

H. $\dfrac{33}{5} \Rightarrow 5\overline{)33.0} = 6.6$

$$\begin{array}{r} 6.6 \\ 5\overline{)33.0} \\ 30 \\ \hline 30 \\ 30 \\ \hline 0 \end{array}$$

Here the division stopped after adding only one zero. There is no remainder.

Calculations are most accurately done using a calculator!

Exercise Set 1 and 2

Add or subtract as indicated. Write your answer under each problem in the space provided. Do not leave any improper fractions, or fractions that can be reduced.

1. $\dfrac{4}{7} + \dfrac{2}{7}$ 2. $\dfrac{5}{8} + \dfrac{3}{8}$ 3. $\dfrac{3}{4} + \dfrac{3}{6}$ 4. $7 + \dfrac{2}{5}$

5. $\dfrac{5}{6} - \dfrac{3}{6}$ 6. $\dfrac{2}{3} - \dfrac{1}{5}$ 7. $5\dfrac{3}{4} - 2\dfrac{1}{4}$ 8. $2\dfrac{1}{4} - 1\dfrac{2}{3}$

9. $\dfrac{15}{24} - \dfrac{3}{8}$ 10. $\dfrac{1}{8} + \dfrac{3}{4}$ 11. $6\dfrac{7}{12} + 2\dfrac{1}{4}$ 12. $9 - \dfrac{3}{5}$

13. $\dfrac{45}{7} - 4$ 14. $5\dfrac{4}{5} + \dfrac{7}{15}$ 15. $\dfrac{3}{10} + \dfrac{1}{20}$ 16. $\dfrac{135}{8} - \dfrac{1}{2}$

Multiply or divide as indicated. Reduce all fractions.

17. $\dfrac{3}{7} \times \dfrac{2}{9}$ 18. $\dfrac{1}{2} \times \dfrac{3}{4}$ 19. $\dfrac{6}{7} \times \dfrac{5}{6}$ 20. $\dfrac{7}{24} \times \dfrac{32}{21}$

21. $5 \times \dfrac{1}{2}$ 22. $\dfrac{4}{7} \times 21$ 23. $\dfrac{1}{4} \times 12$ 24. $5\dfrac{1}{2} \times 3\dfrac{1}{22}$

25. $\dfrac{3}{5} \times \dfrac{5}{3}$ 26. $\dfrac{2}{5} \div \dfrac{3}{4}$ 27. $\dfrac{6}{7} \div \dfrac{6}{8}$ 28. $28 \div \dfrac{4}{5}$

29. $40 \div \dfrac{2}{3}$ 30. $\dfrac{1}{2} \div 4$ 31. $\dfrac{9}{5} \div \dfrac{4}{5}$ 32. $\dfrac{1}{12} \div 36$

Perform the indicated operations and simplify. Round decimal answers to 4 places.

33. $5.289 + 0.4443$ 34. $35.496 - 2.6$ 35. $6 - 0.469$

36. $9 + 0.025$ 37. $3.5 - 2.415$ 38. $14.201 - 5$

39. 0.25×3.402 40. 7×0.009 41. 0.563×0.4

42. 0.405×8 43. 0.001×100 44. 0.01×1000

45. 0.001×10 46. $0.01 \div 100$ 47. $0.1 \div 1000$

48. $6.998 \div 2.07$ 49. $0.025 \div 5$ 50. $369 \div 0.3$

51. $5 \div 6$ 52. $44 \div 0.025$ 53. $22 \div 14$

Calculators that have a fraction key also will convert a mixed number to an improper fraction and/or a decimal. Try the following with and without your calculator. Give three forms of each answer: a. Mixed number b. Improper Fraction c. Decimal.

54. $3\dfrac{2}{5} = ?$ 55. $\dfrac{46}{11} = ?$ 56. 5.75

57. $25\dfrac{1}{5} = ?$ 58. $\dfrac{124}{8} = ?$ 59. 3.25

Make sure to do the parenthesis first.

57. $\left\{ \left(\dfrac{3}{4} \div 3 \right) - \dfrac{1}{4} \right\} + \dfrac{6}{7}$

CHAPTER 3

Algebra Review

In this section you will review how to solve basic equations with one unknown quantity or variable. Remember that an equation has an expression on the left side and an expression on the right side, with an equals sign between.

To solve an equation, you must find a number called a "solution" that will make the left side of the equation equal the right side of the equation when you replace the variable "x" with your number.

Example 1: Determine whether 9 is a solution of the equation $x + 8 = 17$.

$$x + 8 = 17 \qquad \textit{Replace the x with the solution.}$$
$$9 + 8 = 17$$
$$17 = 17 \checkmark$$

Yes, 9 is a solution of the given equation.

Example 2: Determine whether 7 is a solution of the equation $21 \div x = 4$.

$$21 \div x = 4$$
$$21 \div 7 = 4$$
$$3 \neq 4$$

No, 7 is not a solution of the given equation.

SOLVING EQUATIONS INVOLVING ADDITION AND SUBTRACTION

When solving an equation for the variable x, you want to get x alone on one side (usually the left) of the equal sign, and the solution on the other side. In other words, if you end up with **x = 9**, you know that 9 will be a solution to the equation you are solving.

Example 3: Solve the equation $x + 14 = 39$ for x.

Think: What must I do to get rid of the 14 so that I will have x alone on the left?

$$x + 14 - 14 = 39 - 14$$

To get rid of the +14, you must subtract 14 from both sides of the equation to balance the sides.

Step 1: Solve:
$$x + 14 = 39$$
$$x + 14 - 14 = 39 - 14 \quad \textit{Subtract 14 from both sides.}$$
$$x + 0 = 25$$
$$x = 25$$

Step 2: Check:
$$x + 14 \overset{?}{=} 39$$
$$25 + 14 \overset{?}{=} 39 \quad \textit{Replace the x with your solution 25.}$$
$$39 \overset{?}{=} 39 \ \checkmark$$

Example 4: Solve the equation $x - 25 = 106$

Think: What must I do to get rid of the 25 in order to get x by itself?

$$x - 25 + 25 = 106 + 25$$

This time add *25 to both sides to keep the equation balanced.*

Step 1: Solve:
$$x - 25 = 106$$
$$x - 25 + 25 = 106 + 25$$
$$x + 0 = 131$$
$$x = 131$$

Step 2: Check:
$$x - 25 \overset{?}{=} 106$$
$$131 - 25 \overset{?}{=} 106 \quad \textit{Replace the x with your solution 131.}$$
$$106 \overset{?}{=} 106 \ \checkmark$$

Example 5: Solve the equation $57 + x = 75$

Think: What must I do to get rid of the 57?

$$57 - 57 + x = 75 - 57$$

Subtract 57 from both sides to keep the equation balanced.

Step 1: Solve:

$$57 + x = 75$$
$$57 - 57 + x = 75 - 57 \quad \text{\textit{Subtract 57 from both sides.}}$$
$$0 + x = 18$$
$$x = 18$$

Step 2: Check:

$$57 + x \overset{?}{=} 75$$
$$57 + (18) \overset{?}{=} 75 \quad \text{\textit{Replace the x with your solution 18.}}$$
$$75 \overset{?}{=} 75 \ \checkmark$$

SOLVING EQUATIONS INVOLVING MULTIPLICATION AND DIVISION

Keep in mind that $8x$, $8 \cdot x$, $(8)(x)$ all mean 8 times x.

* Note: The times sign is not used in algebra because it would be confusing since an x is often used for the variable.

Example 6:   Solve: $8x = 24$

Think: What must I do to get rid of the 8 that is multiplied by the x?
Divide both sides of the equation by 8, to get x alone.

$$8x \div 8 = 24 \div 8$$

Step 1:   Solve by dividing.

$$8x = 24$$
$$8x \div 8 = 24 \div 8 \quad \text{\textit{Divide both sides by 8.}}$$
$$1x = 3 \qquad\qquad \text{\textit{The 8s on the left side cancel.}}$$
$$x = 3$$

Step 2:   Check:

$$8x \overset{?}{=} 4$$
$$8 \cdot 3 \overset{?}{=} 24$$
$$24 \overset{?}{=} 24 \ \checkmark$$

Example 7:   Solve: $\dfrac{x}{12} = 4$

Think: What must I do to get rid of the 12 that is dividing the x? Multiply both sides of the equation by 12 to get x alone.

$$\frac{x}{12} \cdot 12 = 4 \cdot 12$$

Step 1: Solve by multiplying and reducing.

$$\frac{x}{12} = 4$$

$$\frac{x}{12} \cdot 12 = 4 \cdot 12 \quad \textit{Multiply both sides by 12.}$$

$$\frac{x}{\cancel{12}} \cdot \frac{\cancel{12}}{1} = 4 \cdot 12 \quad \textit{Write 12 as a fraction and cross cancel the 12s.}$$

$$x = 48$$

Step 2: Check:

$$\frac{x}{12} \overset{?}{=} 4$$

$$\frac{48}{12} \overset{?}{=} 4 \qquad \textit{Replace x with your solution.}$$

$$4 \overset{?}{=} 4 \checkmark$$

Example 8: Solve the following equations. ($\cong$ *denotes approximately equal to*)

1.
$$x + 0.45(2.15) = 3.12$$
$$x + 0.9675 = 3.12$$
$$x + 0.9675 - \mathbf{0.9675} = 3.12 - \mathbf{0.9675} \quad \textit{Subtract from both sides.}$$
$$x = 2.1525$$

2.
$$x + 5x = 60$$
$$6x = 360 \qquad \textit{Combine like terms and divide both}$$
$$x = 360 \div \mathbf{6} \qquad \textit{sides by 6. This is \underline{shown} only on}$$
$$x = 60 \qquad \textit{the right side this time.}$$

3.
$$2.3x + 1.5x = \$14.48$$
$$3.8x = \$14.98 \qquad \textit{Combine similar terms}$$
$$3.8x \div \mathbf{3.8} = \$14.98 \div \mathbf{3.8} \quad \textit{divide both sides by 3.8.}$$
$$x \cong \$3.94 \qquad \textit{Round off to nearest cent.}$$

After you are comfortable solving the above equations, you will most likely do the steps on your calculator without writing the steps down. The fact that you add a number to both sides or divide both sides by some constant will be implied by your calculations. Let me demonstrate.

4. Solve:
$$4x + 6 = 86$$
$$4x = 80 \quad \textit{What was done?}$$
$$x = 20 \quad \textit{What was done?}$$

THE DISTRIBUTIVE PROPERTY AND SOLVING EQUATIONS

Although there are many algebraic properties used in solving equations, the one property that needs discussing here, is the distributive property. In Chapter 4 you will be solving percentage equations that rely on the *distributive property*.

$$\text{For any numbers a, b, and c,} \quad \mathbf{a\ (b + c) = ab + ac}$$
$$\mathbf{a\ (b - c) = ab - ac}$$

The distributive property works for both multiplication over addition and for multiplication over subtraction. It also works forward and backward.

Multiplication over addition
$$4(2 + 7) = 4 \cdot 2 + 4 \cdot 7$$
$$4(9) = 8 + 28$$
$$36 = 36$$

Multiplication over subtraction
$$7(9 - 5) = 7 \cdot 9 - 7 \cdot 5$$
$$7(4) = 63 - 35$$
$$28 = 28$$

With a variable over addition
$$(2 + 7)\,x = (2 \cdot x) + (7 \cdot x)$$
$$(9)\,x = 2x + 7x$$
$$9x = 9x$$

With a variable over subtraction
$$(9 - 5)\,x = (9 \cdot x) - (5 \cdot x)$$
$$(4)\,x = 9x - 5x$$
$$4\,x = 4x$$

Note that you can perform the operation within the parenthesis first, or you can "distribute" the constant that is outside the parenthesis to each term inside the parenthesis. Likewise, the x can be on the left of the parenthesis, or it can be on the right of the parenthesis.

$$\mathbf{ab + ac = a\ (b + c)} \quad \Leftrightarrow \quad \mathbf{a\ (b + c) = ab + ac}$$

$$\mathbf{ba + ca = (b + c)a} \quad \Leftrightarrow \quad \mathbf{(b + c)a = ba + ca}$$

When the distributive property (ab + ac = a(b + c)) is used, it is also referred to as factoring out the common factor, which in this case would be a.

Example 9:   Solve: $x + 0.85x = 46.25$

The first term on the left is $1 \cdot x$, while the other is $0.85 \cdot x$. Using the distributive property, we factor out x. $(1 + 0.85)\, x$.

$$x + 0.85x = 46.25$$
$$1x + 0.085x = 46.25$$
$$(1 + 0.85)\, x = 46.25 \qquad \textit{Add inside the parenthesis}$$
$$(1.85)\, x = 46.25 \qquad \textit{divide both sides by 1.85.}$$
$$x = 25$$

Example 10:   Solve: $2.15 + 0.45(2.15) = x$

  Recognize that the 2.15 is a common factor, pull it out.

$$1(2.15) + 0.45(2.15) = x$$
$$2.15\,(1 + 0.45) = x \qquad \textit{ab + ac = a (b+c)}$$
$$2.15\,(1.45) = x \qquad \textit{Add inside the parenthesis.}$$
$$3.1175 = x \qquad \textit{Multiply.}$$

Example 11:   Solve: $315x + 5x = 4640$

The first term on the left is $315 \cdot x$, while the other is $5 \cdot x$. Using the distributive property, we factor out x: $(315 + 5)\, x$.

$$315x + 5x = 4640$$
$$(315 + 5)\, x = 4640 \qquad \textit{Add inside the parenthesis.}$$
$$(320)\, x = 4640 \qquad \textit{Divide both sides by 320.}$$
$$x = 14.5$$

Exercise Set 3

Show each step when solving the equations.

1. Determine if 4 is a solution for the following equations. State yes or no.

    a. $x + 9 = 18$ b. $5 - x = 1$ c. $6x + 4 = 26$

    d. $\dfrac{x}{2} = 2$ e. $(3x - 16) + x = 0$ f. $\dfrac{5x}{6} + \dfrac{2}{3} = 4$

2. Solve these equations. Show all steps and keep your equal signs lined up!

    a. $x + 25 = 49$ b. $42 + x = 106$ c. $21 = x + 6$

    d. $x - 39 = 8$ e. $x - 7 = 150$ f. $4 = x - 14$

    g. $7x = 21$ h. $9x = 108$ i. $24 = 4x$

j. $\dfrac{x}{6} = 3$ k. $\dfrac{x}{5} = 0$ l. $\dfrac{x}{7} = 15$

m. $5x + 2x = 35$ n. $x + 4(2.5) = 12.50$

o. $4x + 8 = 4$ p. $9x - 20 = 7$

3. Using the distributive property, complete and solve. Show your work.

a. $x + 0.45\,x = 43.5$

(____ + ____) $x = 43.5$

(_____) $x = 43.5$

____ $x \div$ ____ = $43.5 \div$ _____

$x =$ _____

b. $x + 0.15\,x = 1.15$

(____ + ____) $x = 1.15$

(_____) $x = 1.15$

_____ $x \div$ ____ = $1.15 \div$ _____

$x =$ _____

c. $533 - 0.25(533) = x$ d. $100(24) - 15(24) = x$

e. $100(x) + 45(x) = \$123.25$ f. $8.95 + 0.25(8.95) = x$

g. $10(x) - 2(x) = 26$ gallons   h. $17(3) + 4(3) = x$

i. $x - 0.39\,x = 402$ j. $25 = 0.12\,(x)$

k. $.08(x) = \$0.08$ l. $42 = x(7)$

UNIT 2

Necessary Math

CHAPTER 4

Percents

I would like to make a prediction! I predict that from this day forward, throughout your long and glorious life, there will never be a day that you do not either hear the word *percent* spoken, see the word *percent* in printed material, or see the % symbol somewhere in your day.

Per-Cent Where do you suppose that term comes from? *Per* means "for each," as in two dinner rolls *per* (for each) customer. You have undoubtedly heard the word *cent*. It could mean a penny, or be short for century (100 years), centennial (every 100 years), or centarian (a person 100 years old). Each of the "c" words used here refers to 100. Now put the *per* and the *cent* together and you get **for each, 100**.

The expression "20 percent of the meat on a turkey is dark meat" means that for each 100 pounds of turkey meat, there will be 20 pounds of dark meat. Said another way "20 pounds of dark meat *for each 100* pounds of turkey," or "20 pounds of dark meat *per 100* pounds of turkey."

Have you ever wondered where the % symbol comes from? Two zeros and a / (slash). The slash represents divided by, and the two zeros represent 100. Therefore, 4% means 4 divided by 100, or 0.04. When you want to convert a percent to a decimal number, just divide by 100 or move the decimal point two places to the left!

To divide by 100, you just move the decimal point two places to the left.

Example 1: Convert the following percents to decimal numbers.

700%	⇒	700	÷	100	=	7
5%	⇒	5	÷	100	=	0.05
32%	⇒	32	÷	100	=	0.32
0.25%	⇒	0.25	÷	100	=	0.0025
9.56 %	⇒	956	÷	100	=	0.0956
2000%	⇒	2000	÷	100	=	20
17.38%	⇒	17.38	÷	100	=	0.1738
0.005%	⇒	0.005	÷	100	=	0.00005

How many pounds of dark meat will there be on a 15-pound turkey? The prior information suggests that there are 20 pounds of dark meat on a 100-pound turkey. Unfortunately this information seems useless, as you have probably never seen a 100-pound turkey at the market! To answer your question, establish a method for finding 20% of any number. That way no matter what size turkey you are cooking, you will know how much dark meat there will be on your bird.

DIRECT PERCENTAGE

To find a percent of any number, you simply multiply. Sound easy? It is. In the language of mathematics, the word **of** translates to **multiply**.

Example 1: Translating from English to mathematics.

"Half **of** the six dozen eggs were broken" .....TRANSLATION.........½ x 6 dozen = 3 dozen eggs were broken.

"I'd like five **of** each kind *of* cookie please. Chocolate chip, macadamia, and oatmeal." ..... TRANSLATION .... 5 x 3 = 15 cookies.

"$\frac{2}{3}$ **of** the 36 people surveyed like sirloin steak better than they like a New York steak." ..... TRANSLATION .... $\frac{2}{3}$ x 36 = 24 people.

Example 2: If a turkey is 20% dark meat, how many pounds of dark meat can you expect from a 15-pound turkey?

Step 1: Convert 20% to a decimal number. $20 \div 100 = 0.20$

Step 2: Multiply 0.20 times 15 pounds. $0.20 \times 15 = 3$

Step 3: State your answer with a sentence.   I can expect 3 lbs of dark meat from a 15-pound turkey.

Example 3: My paycheck last week was $378. State withholding tax is 4%. How much tax was withheld from my paycheck?

Step 1: Convert 4% to a decimal number. $4 \div 100 = 0.04$

Step 2: Multiply. $0.04 \times \$378 = \$15.12$

Step 3: State your answer with a sentence. $15.12 will be withheld from my paycheck.

COMMON STUDENT ERROR

It is important to note that when you hear the words "6 percent," your first reaction should be **OF WHAT?** You should ask yourself the question, 6 percent of what number?

Suppose I have lunch at the Class Act Restaurant, and the check comes to $35 before the tax is added. If the sales tax is 6%, look at the following example and see if you can **find the mistake**?

Subtotal	$35.00
plus 6% tax	.06
TOTAL	$35.06

What error was made? Why? What should the corrected total be? Remember to always ask yourself, 6% of what?

The student mistakenly used the fact that 6% written as a decimal is 0.06, which in dollars and cents is 6¢. The student forgot to ask "6% of what?" As you have figured out already, the tax added should have been *6% of the subtotal* or:

$$6\% \text{ of } \$35.00 \quad \Rightarrow \quad 0.06 \times \$35 = \$2.10$$

making the corrected total $37.10.

MORE PERCENT PROBLEMS

Many other percent problems are used for everyday living, but most can be answered with the following model. This model should be memorized. Learn the words first, then use the symbols.

The **OLD**, plus or minus a percent of the **OLD**, equals the **NEW**.

Note that this model uses parenthesis to indicate multiplication, instead of the common multiplication symbol ×.

$$\textbf{Old} \quad \pm \quad (\%)(\textbf{Old}) \quad = \quad \textbf{New}$$

Example 4: You are given a 16% raise on your salary. Your salary is presently $600 per week, how much will you make after the raise?

*100% of your earnings is currently $600. You will soon get a 16% raise added to your salary. You will then be making **116% of your current salary!** Write an equation to find your new salary.*

$$Old + (\%)(Old) = New$$

$$100\% \ (600) + 16\% \ (600) = \text{New Salary}$$

$$(100\% + 16\%) \ (600) = \text{New Salary}$$

$$116\% \ (600) = \text{New Salary}$$

116% of your current salary will be your new salary.

$$1.16 \ (600) = \text{New Salary}$$

$$696 = \text{New Salary}$$

After your pay raise, you will be making $696.00 per week.

Example 5: One day at work, your boss said "I would like you to mark the broccoli up 45%." You calmly agree and then get your pencil and paper. You note that the price on the a la carte menu is now $2.15 per serving.

Step 1: The **new** price is the price we are looking for. The **old** price is known to be $2.15. The increase is to be 45%. Use the model with a plus sign, because the new price is an increase over the old price.

Step 2: Solve: **Old + (%)(Old) = New**

$$(2.15) + 45\% \ (2.15) = \text{New}$$

$$100\% \ (2.15) + 45\% \ (2.15) = \text{New} \quad \textit{100\% of anything is itself.}$$

$$(100\% + 45\%) \ 2.15 = \text{New} \quad \textit{Use the distributive property.}$$

$$145\% \ (2.15) = \text{New}$$

$$1.45 \ (2.15) = \text{New} \quad \textit{Convert from \% to decimal.}$$

$$3.12 \cong \text{New}$$

Step 3: State your answer. The new menu price for broccoli will be $3.12.

Example 6: The newspaper said that the price of chicken increased by 80% over the last ten years. I pay $1.69 per pound for chicken now. How much was it ten years ago?

Step 1: Note that the **old** price is the price we are looking for. The **new** price is known to be $1.69. The increase is given as 80%. Use the model with a plus sign, because the new price is an increase over the old price.

Step 2: Solve: **Old + (%)(Old) = New**

$$100\% \ (Old) + 80\% \ (Old) \ = \ 1.69 \quad \textit{100\% of old = old.}$$

$$(100\% + 80\%)(Old) \ = \ 1.69 \quad \textit{The distributive property.}$$

$$180\% \ (Old) \ = \ 1.69 \quad \textit{Add.}$$

$$1.80 \ (Old) \ = \ 1.69 \quad \textit{Convert from \% to decimal.}$$

$$Old \ = \ 1.69 \div 1.80 \quad \textit{Divide both sides by 1.80.}$$

$$Old \ \cong \ 0.938$$

Step 3: State your answer: The price of chicken ten years ago was approximately $0.94 per pound.

Example 7: You and a friend are currently renting a two-bedroom condo for which you pay $850 per month. You decide to live alone and move into a one-bedroom condo that rents for 30% less. How much will your new rent be?

Step 1: The **old** rent and the **% decrease** are known. You must find the **new** rent. Use the model with a *minus* sign this time because the new rent is a decrease from the old.

Step 2: Solve: **Old 2 (%)(Old) = New**

$$100\% \ (850) - 30\% \ (850) \ = \ New$$

$$(100\% - 30\%) \ (850) \ = \ New \quad \textit{Use the distributive property.}$$

$$(70\%)(850) \ = \ New$$

$$(0.70)(850) \ = \ New \quad \textit{70\% = 0.70.}$$

$$595 \ = \ New$$

Step 3: I will pay $595 per month rent for my condo unit.

Recall that to change a percent to a decimal number, you divided by 100, or moved the decimal point two places to the left. So it follows that to *change a decimal number into a percent, you simply multiply by 100*, or move the decimal point two places to the right.

An easy way to remember which way to move the decimal point is by thinking about the alphabet. If you begin at the letter D in the alphabet, you must go to the right to get to the letter P. Similarly, to convert a (D) decimal number into a (P) percent, move the decimal point two places to the right. Conversely, to convert a (P) percent into a (D) decimal number, move the decimal point two places to the left. Just as in the alphabet.

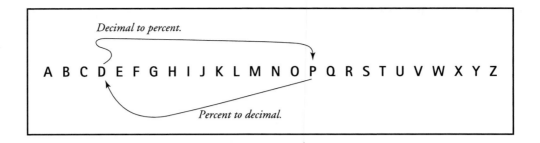

Decimal to percent.

A B C D E F G H I J K L M N O P Q R S T U V W X Y Z

Percent to decimal.

Example 8: The price of cereal has gone up from $4.50 per box to $6.75 per box. Find the percent increase in price.

Step 1: The **old** price and the **new** price are known. You must find the % **increase**. Use the model with a *plus* sign this time.

Step 2: $\text{Old} + (X\%)(\text{Old}) = \text{New}$

$4.50 + (X\%)(4.50) = 6.75$

$(X\%)(4.50) = 6.75 - \textbf{4.50}$ *Subtract 4.50 from both sides.*

$(X\%)(4.50) = 2.25$

$\dfrac{(X\%)(\cancel{4.50})}{\cancel{4.50}} = \dfrac{2.25}{4.50}$ *Divide both sides by 4.50, cancel.*

$X\% = 2.25 \div 4.50$

$X\% = 0.50$ *Convert the decimal to a percent.*

$X = 50\%$

Step 3: The price of cereal has increased by approximately 50%.

The algebraic method to find the % change can be simplified to:

$$\frac{new\ amount - old\ amount}{old\ amount} X\, 100\,\%$$

Observe that a positive number indicates a percent increase, while a negative answer indicates a percent decrease.

Example 9: My dad said that the price of his first calculator was $159. I just bought one for school that cost only $19.95. What a difference! The cost of books has gone sky high, while the price of a calculator has come down significantly. Find the percent decrease in the price of a calculator.

Step 1: The **old** price and the **new** price are known. You must find the **% decrease**. Use the quick model given above.

Step 2: Percent decrease $= \dfrac{new - old}{old} \, x \, 100\%$

$$= \dfrac{19.95 - 159}{159} \, x \, 100\%$$

$$= \dfrac{-139.05}{159} \, x \, 100\%$$

$$= -0.87452 \, x \, 100\%$$

$$\cong -87.45\%$$

Step 3: The price of a calculator has decreased by approximately 87.5%.

Exercise Set 4

Write your answers in the space provided. When an equation is necessary, attach your scratch paper to show your work.

Convert the following percents to decimal numbers, and the decimals to percents.

1. 3%	6. 8.9%	11. 0.006%	16. 0.15
2. 73%	7. 0.45%	12. 0.003	17. 0.4%
3. 345%	8. 23	13. 234.5%	18. 0.2
4. 3	9. 100%	14. 11.73	19. 35.9%
5. 4.6%	10. 2000%	15. 1.6	20. 0.50

21. John pays 29% of his annual income in taxes. If his annual income is $39,000, how much does he pay in taxes?

22. Sandy pays 15% of her annual income in taxes. If her annual income is $28,000, how much does she pay in taxes?

23. A village tax in Illinois is 9% of the total purchases. How much tax is paid on a bill that totals $119.00?

24. A Wisconsin city tax adds 3% to all liquor purchases. How much tax would be paid on a bill that totals $11.44?

25. If a 17% gratuity is added on to your bill, how much gratuity will be added if the bill is $48.75?

26. Sometimes, because people neglect to leave a decent tip, a 15% gratuity is automatically added to the bill. What would the gratuity come to on a bill of $72.00?

27. A good quideline when deciding how much to spend on rent is to spend no more that 30% of your gross monthly income. You earn $1,200 per month. What is the most you should spend on rent?

28. Your food expenses should be no more than 24% of your gross monthly income. Last month you earned $800. Your food purchases totaled $138.78. Did you go over budget?

29. The Sam Choy restaurant grossed $48,000 in one week. Twenty-four percent was spent on labor, and 45% on food costs. How much was spent on labor? How much was spent on food costs?

30. You have decided to mark up everything on the menu by 17%. How much would a club sandwich currently marked $4.79, be on the new menu?

31. After careful thought you have decided to increase your roommate's rent by 12%. Currently you charge her $450 per month. What will her new rent be?

32. The price of gas has decreased by 14% just since last week. If a gallon of gas cost $1.69 today, what was it last week?

33. Due to global warming, the average temperature in Hawaii, in May, is now 76°. What was the average temperature ten years ago if it has increased 2.3% since then?

34. Before Susan started turning off her lights when she left the house, she was paying $65 per month to the electric company. Now she pays only $50 per month. Find the percent decrease.

35. The price of milk increased from $1.15 per half gallon to $1.48. Find the percent increase.

36. The price of artichokes vary throughout the year depending on the season and the weather. In July the price was $ 1.68 each. In December the price soared to $3.60 each. What was the percent increase in price from July to December?

37. Palmer, Alaska has some of the purest water in the world. It is naturally charcoal filtered and 99% pure. How much pure water do you have if you draw 100 gallons from the well?

38. In 1999, some state employees were asked to take a 15% pay cut. Their former annual salary was $65,000. What is their new annual salary?

39. Emma, a college student who graduated in 1972, earned $15,000 per year starting salary. A student graduating in 2003 can expect to earn 320% of Emma's starting salary. How much does the 2003 graduate expect to earn?

CHALLENGE PROBLEMS

40. Lynn lost 25 pounds over the summer, which was a 12% decrease of his original weight. What was Lynn's original weight? How much did Lynn weigh at the end of the summer?

41. There were 1,938 students enrolled at Maui Community College during general registration for the fall semester. By the first day of the semester, the enrollment had risen to 2,575 students. What percent of the total students enrolled late?

42. Amy made $1,365 on an investment. Her broker told her that she had made a clear 15% on her money. How much had Amy invested?

43. A customer left a total of $65 for a meal that costs $54.25. What was the tip rate left for the server?

44. Last year James paid $1,200 for his car insurance. Unfortunately, he had a minor accident in October, which caused his insurance rates to increase by 7%. How much is his new premium?

CHAPTER 5

Fractions and Percents

In Chapter 4 you learned how to convert from a percent to a decimal number, and from a decimal number to a percent. In the food service industry fractions are commonly used instead of decimal numbers to describe measurements, or to use for conversion factors.

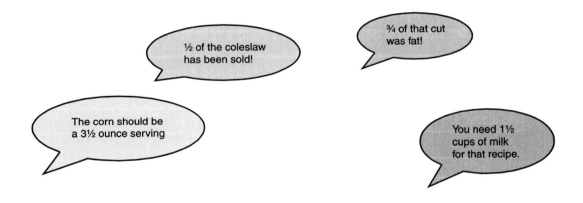

You need to be familiar with the percent names for some common fractions such as:

$$\tfrac{1}{2} = 50\% \qquad \tfrac{1}{4} = 25\% \qquad \tfrac{3}{4} = 75\% \qquad 1 = 100\% \qquad \tfrac{1}{3} = 33\tfrac{1}{3}\%$$

How do you put fractions into percent form? Recall that in the mathematics language a / (slash) means "divided by."

Example 1: Convert 1/8 to a percent.

Step 1: Convert fraction to a decimal number $1 \div 8 = 0.125$

Step 2:   Convert decimal to a percent $0.125 = 12.5\%$

Step 3:   State your answer.   1/8 can be written as 12.5%

Example 2:   Kevin's interest rate is $7\frac{3}{8}\%$ for one year. How much simple interest will he pay on a three-year loan of $1500? *(Recall: Interest = Principal x Rate x Time)*

Step 1:   Convert $7\frac{3}{8}\%$ to a decimal number.

If $(3 \div 8) = 0.375$, then $7 + 3/8 = 7 + 0.375 \Rightarrow 7\frac{3}{8} = 7.375$
so $7\frac{3}{8}\% = 7.375\% = 0.07375$ *Convert percent to a decimal.*

Step 2:   Principal x Rate x Time $1500 \times 0.07375 \times 3 = 331.875$

Step 3:   State your answer with a sentence.   He will pay $331.88 interest on the loan.

Example 3:   Harris put $500 in a special savings account and was told that he would earn $8\frac{1}{4}\%$ interest at the end of one year, providing he did not withdraw any of his money before the year was over. How much was in Harris' account at the end of the year? *(after the interest was added to his original deposit)*

Step 1:   We could first find the interest earned using the PxRxT formula, then add that amount to the original principal to find our total . . . OR do both steps in one using the model from Chapter 4.

$$Old \pm (\%)(Old) = New$$

$$500 + (8\frac{1}{4}\%)(500) = New$$

$$1(500) + 0.0825(500) = New \quad \textit{Convert } 8\frac{1}{4}\% \textit{ to a decimal number.}$$

$$1.0825(500) = New$$

$$541.25 = New$$

Step 2:   State your answer: After one years, Harris will have $541.25 in his account.

> **Note:** In Example 2, we found the interest only.
> In Example 3, we found the amount in the
> account after the interest had been added.

PART TO THE WHOLE $\dfrac{part}{whole}$

"Three out of four people prefer rice over baked potatoes." These words are comparing people who prefer rice with people who prefer baked potatoes. In mathematical terms *three out of four* can be written as ¾, 3 ÷ 4, or 3 : 4. All of these are acceptable ways of showing comparisons, or writing ratios.

   A *ratio* is a comparison of two quantities expressed as a quotient of the first divided by the second. Fractions are ratios used to compare equal parts of a whole.

$\dfrac{3}{4}$ means 3 of 4 equal pieces

Let's play Jeopardy!   The answer is "45% of this class is male." What is the question?

♫ ♪ ♫ ♪ . . . Time's up. Your answer please?

"What part of this whole class is male?"

Right! You are comparing a **part** of the whole class *to the* **whole** class.

Example 4:   A side of beef weighs 350 lbs, the flank weighs 18 lbs. What percent of the side of beef is flank?

Step 1:   Define the part and the whole: 18 lbs is the part and
                                              350 lbs is the whole side

Step 2:   Show the $\dfrac{part}{whole}$ $\dfrac{18}{350}$

Step 3:   Divide and convert to % $18 \div 350 = 0.051 \Rightarrow 5.1\%$

Step 4:   State your answer.   5.1% of the side of beef is flank.

Some of you are very comfortable with algebra, so here is another way of solving this type of percent problem using algebra.

Referring to Example 4: What percent of the side of beef is flank?

ALGEBRAIC TRANSLATION . . .

$$X \% \times 350 = 18$$

$$X\% = 18 \div 350$$

$$X\% = 0.051$$

$$X = 5.1\%$$

Therefore 5.1% of the side of beef is flank.

Exercise Set 5

Convert the following fractions to percents. Round all decimals to two places.

1. $\dfrac{3}{4}$

2. $\dfrac{5}{6}$

3. $\dfrac{3}{20}$

4. $\dfrac{1}{8}$

5. $\dfrac{3}{5}$

6. $\dfrac{11}{6}$

7. $\dfrac{1}{6}$

8. $\dfrac{4}{15}$

9. $\dfrac{1}{2}$

10. $\dfrac{1}{4}$

11. $\dfrac{4}{50}$

12. $\dfrac{69}{46}$

13. $\dfrac{1}{75}$

14. $\dfrac{8}{7}$

15. $\dfrac{1}{3}$

16. Gail invested $10,000 in her sister's home. She was to receive 8¼% simple interest for 5 years. How much total interest did she earn? (hint: I = P × R × T)

17. Delia borrowed $8,000 against her life insurance policy. The interest rate was 8¼% per year. How much interest did she owe at the end of one year?

18. Eric put $5,000 in a money market certificate earning 6⅗% interest annually. How much is in the account at the end of one year?

19. There are 15 males and 13 females in your math class. a) What percent are females? b) What percent are males?

20. Your brother wants 11½ % simple interest on a loan he gave you. You borrowed $885.75 from him. How much interest will he expect?

21. Twelve of the shelter's cats are spayed. There are 18 cats altogether. What percent of cats still need to be spayed?

22. Tell what percent of the whole side of beef each butcher's cut makes.

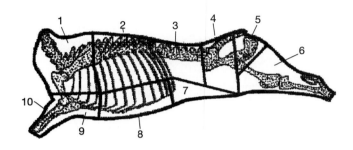

Total weight _____?_____

- 1 = chuck, 84 lbs.

- 2 = rib, 37½ lbs.

- 3 = short loin, 16 lbs.

- 4 = sirloin, 20 lbs.

- 5 = rump, 25¼ lbs.

- 6 = round, 72 lbs.

- 7 = flank, 18 lbs.

- 8 = plate, 23 lbs.

- 9 = brisket, 24 lbs.

- 10 = shank, 15½ lbs.

23. Tell what percent of the side of pork each Butcher cut represents.

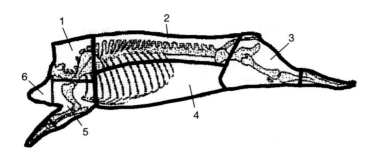

Total weight ___?___

- 1 = blade bone and spare rib, 28 lbs.

- 2 = loin, 40 lbs.

- 3 = ham, 20 lbs.

- 4 = belly, 47 lbs.

- 5 = butt and picnic, 26 lbs.

- 6 = jowl, 11 lbs.

CHAPTER 6

Interest: Simple, Compound, and Credit Card

In Chapter 5, you were asked you to recall the simple interest formula $\mathbf{I = P \times R \times T}$. The interest you must pay on a loan, or the interest that is paid to you on an investment, is equal to the **P**rincipal (the amount you have borrowed or invested) times the **R**ate (the percent mark up on your loan or investment) times the **T**ime (the amount of time the money is either borrowed or invested).

When using the simple interest formula, keep in mind that the rate and the time must be written in the same units. For example, 4% per *year* for 13 *years*, or 2% per *month* for 18 *months*. Notice the units are consistent. However, if given an interest rate of 6% per *year* for 18 *months*, you should begin by converting the 18 months to 1.5 years.

Remember that unless otherwise stated, interest rates are assumed to be ***per year***.

Example 1: Teresa borrowed $1500 to buy a computer for her business. The loan was for three years at an interest rate of 9% per year. Figure her simple interest.

Step 1: The formula needed is I = P x R x T where: P = $1,500
    R = 9% or 0.09
    T = 3 years

Step 2: Calculate   I = P x R x T
    I = 1500 x 0.09 x 3
    I = 405

Step 3: State your answer: Teresa had to pay $405 interest on her loan.

Example 2: How much simple interest will Jake have to pay on a loan of $2,400 if the rate is 14.5% per year and he borrowed the money for 16 months?

Step 1: The formula needed is I = P x R x T where

    P = $2,400   *Given.*

    R = 14.5% $\Rightarrow$ 0.145   *Convert percent to decimal number.*

    T = 16 months $\Rightarrow \dfrac{16}{12} = \dfrac{4}{3}$ *Convert months to years.*

Step 2:   Solve: I = P x R x T

$$I = 2400 \times 0.145 \times \frac{4}{3}$$

$$I = 464$$

Step 3:   State your answer: Jake will pay $464 in interest on his loan.

COMPOUND INTEREST

Most of you realize that there is a difference between simple interest and compound interest. Compound interest is often described as **paying interest on interest**. It is most often used on investment-type savings accounts. To understand compound interest it is best to work out a few problems without the standard formula.

Compound interest is based on the idea that after you deposit an amount into your savings account, you do not withdraw any of the money. The bank pays you interest on your original deposit the first time, then on your original deposit plus the interest they already paid you, the second time. The third time is much the same. The bank just continues to pay interest on the amount that accumulates in your untouched account. If that seemed like a mouthful, let me demonstrate.

We will use the model that you memorized in Chapter 4 with a slight alteration for **Time**.

(Old) + (%)(Old) = New   ⇒   Principal + (Principal)(Rate)(Time) = New

---

P × R × T	You may notice that this part of the equation calculates the simple interest, which, when added to the original principal, gives the new principal. (*The order doesn't matter when multiplying.*)

Example 3: Brandon deposited $500 in a savings account that advertised 8 3/4% interest compounded semiannually. At the end of six months, the bank figured interest on Brandon's savings and directly deposited it into his account. (*Semiannually* means twice per year, *quarterly* means four times per year, *bimonthly* means every two months.)

Principal + (Principal)(Rate)(Time) = New

(Old) + (%)(Old)(Time) = New   *For 6 months we use*

500 + (8¾%)(500)(½) = New   *1/2 of the 8 3/4 % yearly rate.*

500 + (0.0875)(500)(0.5) = New   *8 3/4 % = 0.0875.*

500 + 21.875 = New

521.875 = New

Brandon now has $521.88 in his savings account. After six more months, the bank once again adds interest to Brandon's account using his new balance as the principal.

$$\text{Principal} + (\text{Principal})(\text{Rate})(\text{Time}) = \text{New}$$

$$\text{Principal} + (\text{P x R x T}) = \text{New}$$

$$521.875 + (8\tfrac{3}{4}\%)(521.875)(1/2) = \text{New} \qquad \textit{Again, ½ of the yearly rate.}$$

$$521.875 + (0.0875)(521.875)(0.5) = \text{New}$$

$$521.875 + 22.83 = \text{New}$$

$$544.71 \cong \text{New}$$

So after one year, Brandon has $544.71 in his savings account. Let's try it one more time, and then I will assume you can continue with this pattern endlessly.

$$\text{Principal} + (\text{P x R x T}) = \text{New}$$

$$544.707 + (8\tfrac{3}{4}\%)(544.707)(1/2) = \text{New} \qquad \textit{1/2 of the yearly interest.}$$

$$544.707 + .0875(544.707)(0.5) = \text{New}$$

$$544.71 + 23.83 = \text{New}$$

$$568.54 \cong \text{New}$$

And finally after 18 months, or three semiannual interest payments, Brandon has a total of $568.54 in his savings account.

Here is the **Compound Interest Formula.**

$$A = p\left(1 + \frac{r}{n}\right)^{nt}$$

A = amount, p = principal r = annual rate n = periods per year t = time in years

---

For this next example, let's compare methods: the long method on the left and the formula on the right. Keep in mind that in order to have both methods come out exactly the same, we could not round off any of the decimal numbers. However, here I have chosen to round each new principal off to three decimal places. Compare the difference in the final outcome.

Example 4: Samantha bought a certificate of deposit in the amount of $5,500. The interest rate was 6.45% compounded quarterly if the money was not withdrawn for one full year. How much was Samantha's CD worth at the end of one year? How much interest did Samantha earn?

(Remember quarterly means 1/4 of the year. So the interest quarterly is 1/4 of 6.45% or approximately 1.6125%, which is 0.016125 in decimal form!)

After 3 months

P + (P x R x T) = New
5500 + (1/4)(6.45%)(5500) = New
1(5500) + 0.016125(5500) = New
1.016125(5500) = New

5588.688 ≅ New

After 6 months

P + (P x R x T) = New
5588.688 + (1/4)(6.45%)(5588.688) = New
1(5588.688) + 0.016125(5588.688) = New
1.016125(5588.688) = New
5678.805594 = New

5678.806 ≅ New

After 9 months

P + (P x R x T) = New
5678.806 + (1/4)(6.45%)(5678.806) = New
1(5678.806) + 0.016125(5678.806) = New
1.016125(5678.806) = New
5770.376747 = New

5770.377 ≅ New

After one year

P + (P x R x T) = New
5770.377 + (1/4)(6.45%)(5770.377) = New
5770.377 + 1.016125(5770.377) = New
1(5770.377) + 0.016125(5770.377) = New
1.016125(5770.377) = New
5863.424329 = New

$ 5863.42 ≅ New

Using the Compound Interest Formula

Using a calculator you will be working inside the parenthesis first.

$$A = p\left(1 + \frac{r}{n}\right)^{nt}$$

p is $5500, r is 6.45%, $n = 4$ periods per year and $t = 1$ year

We will do the $r \div n$ first, then add the 1.

$$\left(1 + \frac{r}{n}\right)$$

press 0.0645 ÷ 4 = 0.016125
add the 1 and get 1.016125

Next, we take the result from the parenthesis 1.016125, and raise it to the power n x t

$$\left(1 + \frac{r}{n}\right)^{nt}$$

(1.016125) ^ (4 × 1) = 1.066076932

Lastly, multiply by the principal amount p.

$$p\left(1 + \frac{r}{n}\right)^{nt}$$

5500 × 1.066076932 = 5863.423126

And finally, **$A \cong \$5,863.42$**

⇒ Samantha's certificate of deposit was worth $5,863.42 after one year.

⇒ Her total interest earned came to $ 363.42   *(Subtracting: $5863.42 – $5500)*

CREDIT CARDS (OPTIONAL)

No discussion about compound interest would be complete without taking the time to talk about how credit card companies figure their finance charges. Usually, the **disclosure** part of your credit card statement letting you know how finance charges **accrue** is so complicated, and laced with words that only people in the banking industry would understand, that most of us just sigh and think, "What are they talking about?" Here is a perfect example.

Important Information about your United College Plus First Card Account
1. You may compute the FINANCE CHARGE (due to Periodic Rates)
    a) By multiplying each of the Corresponding Finance Charge Balances (shown on the front) by the number of days in the billing period, and
    b) Then multiplying each of the results at (a) by the applicable daily periodic rate, and
    c) Then adding each of the products from the computation at (b) together.

The daily balance of purchases, advances, Fees and Finance Charges (on which Finance Charges at each applicable periodic rate are computed) are determined separately by taking the beginning balance of purchases, advances, Fees and Finance Charges, adding any new purchases, advances, Fees and Finance Charges that accrued or were posted that day, and subtracting any payments or credits applied that day to such purchases, advances, Fees and Finance Charges. The Corresponding Finance Charge Balances are averages of the daily balances for the billing period and, except as provided in paragraph 2, include:
- Purchases, advances and Fees beginning on the date of posting up to the date payment is posted by us, and
- Finance Charges(Due to Periodic Rates) beginning on the day after the date they were accrued up to the date when payment is posted by us.

Let's begin with some definitions. According to an American Heritage Dictionary, *accrue* means "to come to one as a gain, and addition, or an increment; interest accruing in my savings account." For the purposes of your credit card company, accrue means to add up or accumulate finance charges.

Next we have **APR**, which stands for the Annual Percentage Rate. "The CURRENT DAILY PERIODIC RATE, which is the APR divided by 365 days, is calculated by taking the highest rate among the prime rates(s) published in *The Wall Street Journal* on the first day of the month in which the beginning day of the billing cycle occurs, rounded to the nearest 1/10 of a percent then adding the margin or "spread" indicated by your card carrier (from 1.4% to 9.9%) to that rate."

. . . Good grief, what a mouthful !!!

Billing Cycle: This is the 28 to 31 days included in your monthly billing statement: April 15th to May 12th would be a 28-day billing period.

Daily Purchase Balance: Periodic Rate *finance charges* accrue on a daily basis. "To determine the Daily Purchase Balance, we take the beginning purchase balance, including accrued but unpaid *finance charges*, each day, subtract any payments or credits, and add new purchases and all other charges. This gives us the daily purchase balance."

Now let's try! Apply the *current daily periodic rate* to the *daily purchase balance* to see how interest is *accrued* for one day on a GoldPlus® credit card.

Example 5: The current daily periodic rate is 0.08673%, and your daily purchase balance on October 17, 2000 is $1,300. Find the finance charges that accrue for that day and tell your opening balance on October 18, 2000.

Step 1: This is a direct percent problem. To find the finance charge, you must take 0.08673% of $1,300. To find the next day's opening balance, just add the original balance plus the finance charge.

Step 2: Calculate finance charge first:

$$0.08673\% \text{ of } \$1,300 = \text{Finance Charge}$$

$$0.0008673 \times \$1,300 = \text{Finance Charge} \quad \textit{Convert \% to decimal.}$$

$$\$1.12749 = \text{Finance Charge}$$

Step 3: Now add the original balance plus the finance charge:

$$\$1,300 + \$1.12749 = \$1,301.12749 \quad \textit{Original + finance charge.}$$

If you look closely, this is a case of the old + (% • old) = new *where the finance charge is the (% of the old)!*

Step 4: State you answer:

The finance charge accrued on 10/17/2000 was $1.13. This brought your opening balance on 10/18/2000 to $1,301.13.

The following excerpt was taken from a "Cardmember Agreement Notice"

NOTICE OF CHANGE IN TERMS TO YOUR FIRST CARD® CARDMEMBER AGREEMENT:

A.   (1) Finance Charges (Due to Periodic Rates) will be imposed each day (a) on purchases advances and Fees beginning on the day they were posted to your account and (b) on Finance Charges beginning on the day after the date they were accrued in your account. Section 7.A.(9) below describes how you can avoid certain Finance Charges (Due to Periodic Rates). Purchases and advances (other than First Card Checks) are posted on the later of the transaction date or the first day of the billing period in which the transaction is billed to your account. First card checks are not posted until we receive evidence that the advance has been made.

Look very closely at the part that says *"(1) Finance charges . . . will be imposed each day . . . (b) on Finance Charges . . ."* Like compound interest, you are paying interest charges on the interest that you were charged the previous day. That is to say that your daily finance charge is based on your new daily balance, which includes the previous day's interest charges. Take a look.

Example 6: On 10/18/2000, referencing the account in Example 5 above, your beginning purchase balance was $1,301.13, which included the finance charge from the previous day. This day, three new purchases that you had made cleared the bank. They were in the amounts of $500.26, $45.00, and $76.98. The Daily Periodic Rate on 10/18/00 was 0.08921%.

    a.   Find the finance charges imposed that day, and tell your opening balance on 10/19/2000.

    b.   What was the total finance charge accrued for 10/17 and 10/18?

a) To find the finance charge, you must take 0.08921% of the Daily Purchase Balance, which is sum of the new purchases *plus* the previous day's balance. *Thus adding interest to the previous day's interest, compound interest, as well as to your past and present purchases.*

Step 1: Find the Daily Purchase Balance, which is:

new purchases + previous day's balance

i) new purchases:

$$\$500.26 + \$45 + \$76.98 = \$622.24$$

ii) add the previous day's balance:

$$\$622.24 + \$1,301.13 = \$1,923.37$$

Step 2: Multiply by the daily % rate to find the finance charges:

$$\$1,923.37 \times 0.0008921 = \$1.716$$

Step 3: Find opening balance on 10/19:

$$\$1,923.37 + \$1.716 = \$1,925.086$$

Step 4: State the answer:

The finance charge for 10/18/00 was $1.72 and the opening balance on 10/19 will be $1925.09.

b) Add the finance charges on 10/17 and 10/18:

$$\$1.13 + \$1.72 = \$2.85 \text{ total finance charges}$$

$\Rightarrow$ The total finance charges accrued on 10/17 and 10/18 was $2.85.

Example 7: Find the **Current Daily Periodic Rate** for an account whose margin or spread is 9.8%. Given that the highest prime interest rate index, according to the *Wall Street Journal*, was based on an 8.36% APR.

Step 1:   The Current Daily Periodic Rate is the APR (rounded to the nearest tenth percent) <u>plus</u> the given margin or spread of your credit card company. This sum is then divided by 365 days.

Step 2: Calculate

$$\overset{\textit{margin}}{(9.8\%} + \overset{\textit{APR}}{8.4\%)} \div 365 \ = \ \textbf{Current Daily Periodic Rate}$$

$$18.2 \ \% \div 365 \ = \ \textbf{Current Daily Periodic Rate}$$

$$0.049863 \ \% \ = \ \textbf{Current Daily Periodic Rate}$$

Step 3: State your answer: The **Current Daily Periodic Rate** is 0.049863 %.

THE GOOD NEWS IS:

Most credit card companies will give you a break on your finance charges if you pay your balance in full every month. Here are two such excerpts:

"In calculating daily finance charge balances for a billing cycle, we will (a) exclude purchases and fees posted in the billing cycle if you fully paid your balance for the previous billing cycle, and b) exclude purchases, finance charges due to periodic rates, and fees if you fully paid your balance for the previous two billing cycles." (First Card ®)

"A Finance Charge is imposed on a Purchase form the date the Purchase is included in the Daily Balance until the date payment in full is received. However, no Finance Charge is imposed on new Purchases in the Billing Cycle in which they are posted to your Account if in the prior Billing Cycle you had no New Balance, or if you paid the entire New Balance on your Account by the payment Due Date on the monthly statement for that cycle." (AT&T Universal Card®)

In other words, if you pay your credit card bill in full and it arrives at the credit card company by the due date, there will be no finance charge.

 The moral to this story is to pay your credit card balance in full each month, before the due date.

Exercise Set 6

Round to the nearest cent.

1. What is the interest on a $1500 principal at the rate of 16% for one year?

2. Find the interest for 18 months on a $2300 principal, given an 8.5% annual rate.

3. Find the simple interest on a principal of $3500 at 7% for 3 years.

4. Find the simple interest paid on a principal of $6500 at 63/4% for 5 years.

For problems 5 – 12, find the simple interest.

	Principal	Annual Interest Rate	Time	Simple Interest
5.	$350	8%	2 year	
6.	$6,740	11.4%	6 years	
7.	$850	6.35%	9 months (hint: ¾ year)	
8.	$2,500	16½ %	15 months (hint: ¹⁵⁄₁₂ of a year)	
9.	$770	9%	18 months	
10.	$5,000	6%	36 months	

11. $2,100   10%   3 years

12. $972.45   8%   1 year

13.   Carlyle deposited $2,432 in a savings account that was giving 6½% interest annual interest, compounded semiannually. Find the amount in Carlyle's account at the end of two years, assuming he withdrew nothing until then.

14.   Johanna opened a savings account at the Las Cruces Savings and Loan with a $5,000 deposit. The advertised interest rate was 7¼% compounded semiannually. How much money was in Johanna's account after 18 months if she held off withdrawing even a penny until then?

15.   In his dreams, Steve opened a bank account with a deposit of $7,560. The dream bank paid an annual interest rate of 12% compounded quarterly. He left his money in the bank for 15 months.

       a) How much was in Steve's account after 15 months?

       b) How much interest had he earned?

16.   Tracy managed to save $250 before his tenth birthday. He put it in a savings account that earned 6% annual interest, compounded quarterly, if the money was left in the account for 3 full years.

       a) How much interest had Tracy earned in 1 year?

b) How much was in Tracy's account at the end of 1 year?

c) How much was in Tracy's account after 3 years?

17. Find the new principal and the total interest paid at the end of 18 months on an investment of $3,400. The investment rate is 4%, compounded semiannually.

18. Find the amount of interest for one year: Principal: $43,000, Interest Rate: 11% compounded semiannually.

19. Derrick borrowed $8,000 from his parents to buy a sailboat. They required him to make payments once per year. His first payment was to be $2,500 plus 4% simple interest on the remaining balance. How much interest did Derrick pay to his parents that first year?

20. Last year Kiana borrowed $1,800 from her parents to attend Maui Community College. Her parents agreed that if Kiana got straight As she would not have to pay the loan back. Unfortunately, Kiana got a B in Math. She agreed to pay her parents 5% simple interest on the whole amount. How much interest did Kiana pay after one year?

(OPTIONAL)

21. Bailey ran her *Visa*® credit card bill up to $1,752.50 in one day. Figure the finance charges for that day, if the daily periodic rate is 0.0456%.

22. If the Daily Purchase Balance on your MasterCard® is $687.29, figure the finance charge for that day, given that the current daily periodic rate is 0.0399%.

23. If the APR for a given day is 9.996% and the margin or "spread" for a particular credit card company is 8.5%:

    a) Find the monthly periodic rate. (*Hint: divide by 12*)

    b) Find the daily periodic rate.

24. Find the daily periodic rate, for your Visa® credit card if the margin for that card is 9.3% and the APR according to The *Wall Street Journal* is 11.432%.

25. On May 5 the Daily Purchase Balance on your *Discover Card*® was $456.17. On May 6 two new charges of $36.78 and $112.34 were posted. If the Daily Periodic Rate for those days was 0.0682%, find your opening balance on May 7.

Challenge Problem

26. Assume that for problem 25 above you make no more purchases until the end of your billing cycle, May 10. What would be the amount due on your statement? *(end of the day on May 10)*

"How We Compute the Finance Charge: Finance Charges on your Account for Purchases are calculated by applying a Daily Periodic Rate (nominal Annual Percentage rate divided by 365 days) to your Daily Balance of Purchase and adding together any such Finance Charges for each day in the Billing Cycle." AT&T Universal Card®

CHAPTER 7

Pie Graphs and Bar Graphs

Look at the following information and determine which presentation you prefer, the paragraph, the table, or the pie graph.

Paragraph The percent of each primal cut of pork taken from a 148-pound side of pork is given. The leg is 21 lbs or 14.18%, the belly is 40 lbs or 27.02%, the loin is 30 lbs or 20.27%, the hand and spring are 12 lbs or 8.10%, the blade bone is 22 lbs or 14.86%, the sparerib is 14 lbs or 9.45%, and the head is 9 lbs or 6.08%.

Table

Cut	Pounds	Percent
Leg	21	14.19%
Belly	40	27.03%
Loin	30	20.27%
Hand/Spring	12	8.11%
Blade Bone	22	14.86%
Sparerib	14	9.46%
Head	9	6.08%
Total	148	100.00%

3D Pie Graph

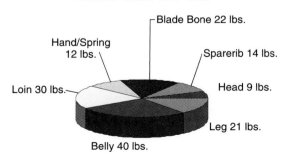

Primal Cuts of Pork

PIE GRAPHS

A *pie graph* is used to illustrate how one whole group is broken down into subgroups. It shows visually what part or size of the whole each subgroup represents. Keep in mind that the total of all wedges should equal 100% of the whole group.

A pie graph is made using a circle, which you may recall has 360°. To make a pie graph you need a compass or any round object such as a glass that you can trace around, a ruler, and a protractor.

Let's start by making a simple pie graph that requires no calculations.

Example 1: "You may distribute the shipment of napkins equally among the main dining room, the pool bar, the terrace buffet, and room service," the boss said.

Step 1: How many subgroups are required?　　　　4 wedges are needed.

Step 2: What % of the pie will each wedge represent?　　　1 pie ÷ 4 wedges
　　　　　　　　　　　　　　　　　　　　　　　　　　　¼ = 25%.

Step 3: Draw a circle, divide it into wedges.　　　　Divide the pie into 4 equal parts.

Step 4: Label each wedge.　　　　Name the pieces.

Step 5: Label the pie graph.　　　　Distribution of Napkin Shipment.

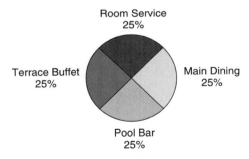

Distribution of Napkin Shipment

Room Service 25%

Terrace Buffet 25%

Main Dining 25%

Pool Bar 25%

World Famous Hula Pie made with Macadamia nut ice cream and lots of hot fudge, begins with an Oreo cookie crust . . . yummy!

Example 2: Last year Hula Pie was our best-selling dessert. We sold 2,535 servings. Of those servings, 35% were sold at the Pool Bar, 28% at the Terrace Buffet, 15% through room service, and the rest at the main dining room. Create a pie graph to illustrate the distribution of sales.

Step 1: How many subgroups are required?   4 wedges are needed

Step 2: What % of the pie will each wedge represent?

To answer that question, you must first recall that the sum total of all wedges must equal 100% of the Hula Pie sales. You were given the percent for each service area except the Main Dining Room. You must calculate what percent of the total sales of Hula Pie was sold in the Main Dining Room.

	Pool	35%
	Terrace	28%
+	Room Service	15%
	Main Dining Room	? %
	Total	100%

Subtract to find the main dining room percent.

$$100\% - (35\% + 28\% + 15\%) = ? \%$$

Main Dining Percent of Sales = 22%

Step 3: How big should each wedge be? Find the wedge sizes in degrees.

- Pool Bar   35% of 360° $\Rightarrow$ 0.35 x 360° = 126°

- Terrace Buffet   28% of 360° $\Rightarrow$ 0.28 x 360° = 100.8°

- Room service   15% of 360° $\Rightarrow$ 0.15 x 360° = 54°

- Main dining   22% of 360° $\Rightarrow$ 0.22 x 360° = 79.2°

Step 4: Draw the circle and use a protractor to measure the degrees for each wedge.

Hula Pie Sales Distribution

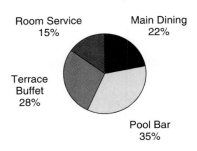

Room Service
15%

Main Dining
22%

Terrace
Buffet
28%

Pool Bar
35%

Example 3: Use the information given in Example 2 to find out how many servings of Hula Pie were sold in each of the four areas.

Step 1: First find out how many **total** Hula Pie servings were sold?

Given: 2,535 total servings

Step 2: You know what percent of the Hula Pie servings were sold in each area, now ask yourself **percent of what**? Yes! Percent of the total servings sold!

		Servings		% of Total
Pool Bar	35% x 2,535 $\Rightarrow$	0.35 x 2535 = 887.25	$\cong$ 887
Terrace Buffet	28% x 2,535 $\Rightarrow$	0.35 x 2535 = 709.80	$\cong$ 710
Room Service	15% x 2,535 $\Rightarrow$	0.35 x 2535 = 380.25	$\cong$ 380
Main Dining	22% x 2,535 $\Rightarrow$	0.35 x 2535 = 557.70	$\cong$ 558

* *Note:* $\cong$ *stands for "approximately equal to"*

Step 3: There were 887 servings sold in the Pool Bar, 710 sold at the Terrace Buffet; 380 were delivered by room service and 558 were consumed in the Main Dining Room. (*Notice that the percentages were rounded to the nearest whole number as only whole portions are served.*)

BAR GRAPHS

Although a **bar graph** is not often used to show percents of a whole, it is an excellent way to show comparisons. For instance, if you wanted to compare supermarket prices of milk over a one-year period, a bar graph would do nicely. You may make the bars vertical or horizontal, so that's the first thing you need to decide.

Next, it is very important to choose a good **scale** for your numerical comparison. If you are comparing prices of milk, you wouldn't want to number from $0 to $100.00 on the price scale, right? Why not?

You must also decide how you want to break down the year: Daily? Weekly? Monthly? Bimonthly? Seasonally? If you are putting your graph in a very small space, you may want to set up the horizontal axis as a seasonal scale, thus requiring only four divisions.

The *amount of information* that you want to convey to the reader, as well as the **readability** of the graph, is very important. These two factors, along with your *given space*, will determine how you number and label your scales.

If you choose vertical bars, then your horizontal scale should represents the "independent quantity," while the vertical scale is reserved for the "dependent quantity." However, if you choose horizontal bars, the independent and dependent scales are reversed as in Example 5.

How do you know which quantity is dependent and which is independent? Suppose you want to make a graph to compare *monthly* milk *prices* for a year. Which quantity, month or price, depends on the other? The price depends on the month.

Example 4: Draw a bar graph to illustrate the price comparison for milk for in 1998. Use the information taken from the government statistics website. Prices Received by Farmers, Milk, US 1998

Note: the prices are per *100 weight* (≅12.5 gallons). See exercise set for a complete explanation of cwt or 100 weight.

January	$14.70
February	$12.90
March	$14.40
April	$14.00
May	$13.30
June	$14.00
July	$14.10
August	$15.40
September	$16.60
October	$17.60
November	$17.90
December	$18.00

Step 1: Because I do have enough space for my milk price comparison graph, I will let the horizontal axis represent each month of the year, while the vertical axis is marked off in $1.00 increments, beginning with $10.

Price Paid to Farmers for Milk in 1998

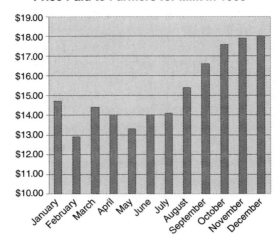

Example 5: Draw a bar graph to enhance the findings from Example 2.

Hula Pie Sales

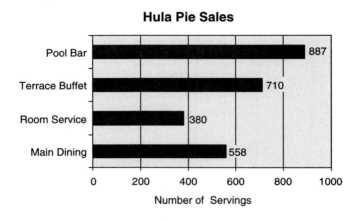

Example 6: Look at the following bar graph. Discuss what you see.

%Anchorage Costs Above U.S.
1976–1995

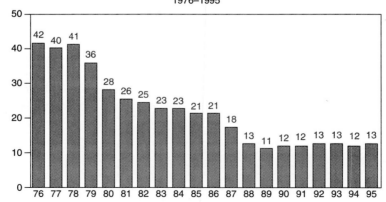

Source: 1975–81 based on *Urban Intermediate Budget for a 4-Person Family. By U.S. Department of Labor, Bureau of Labor Statistics. Estimates for 1982–1995 inflate the 1981 data using annual changes in the Anchorage & U.S. CPI.

1. What exactly is the graph showing the viewer?

    The cost of living in Anchorage, Alaska, appears to be higher than in the rest of the United States. "How much higher?" is the purpose of the graph. The graph is based on an Urban Intermediate Budget for a four-person family and compares, by percentage, the cost of living in Alaska to the cost of living in the lower 48 states. (*The lower 48 refers to all of the United States except Hawaii and Alaska.*)

2. If a "lower 48" family of four spent $15,000 on living expenses in 1980, how much did a comparable family living in Alaska spend?

    The graph tells us that in 1980, a family living in Alaska spent 28% more on living expenses than did a family in the Lower 48 states.

    $$\text{Old} + (\%) \text{ of (Old)} = \text{New}$$

    "Lower 48" Expenses + (18%)("Lower 48" Expenses) = Alaska Family Expenses

    15,000 + (0.28)(15,000) = Alaska Family Cost of Living

    1.28(15,000) = Alaska Family Cost of Living

    $19,200 = Alaska Family Cost of Living

Conclusion: In 1980, an Alaskan Family of four spent approximately $19,200 on living expenses, while a comparable "lower 48" family spent only $15,000.

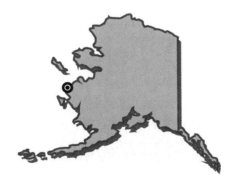

Exercise Set 7

1. What is a pie graph used for?

2. How is a bar graph best used?

3. Shelly wants to compare how many hours of study for the last exam each student in her class undertook. What type of graph should Shelly use?

4. Mead wants to make a graph to show his students what percent of the class earned As, Bs, Cs, Ds, and Fs on the last exam. What type of graph should he use?

5. Why use any graph at all? What can a graph do that an informative paragraph can't?

6. How often and where do you see examples of graphs?

7. Cut out an example of a bar graph and a pie graph from a newspaper or magazine. Discuss what the graph represents. Is it reader friendly? Is it well labeled? Are the colors used well?

8. "Ralph, please distribute these menus evenly among our eight restaurants."

    a. What percent of the menus will each restaurant get? Show your work.

b. If you draw a pie graph to represent this request, how would you divide the circle? Would you need a protractor?

9. Ms. Tanaka wants to illustrate how much students spent in the cafeteria this past year, as compared with student spending over each of the prior three years.

    a. Tell what her horizontal and vertical axis would represent if she used a bar graph.

    b. What would be the title of the graph?

    c. Can a pie graph be used? What would each wedge represent? What would the whole pie represent?

Unit 2 Summary—fill in the Definitions and Explanations for Review

Converting Percent to Decimal $\quad$ $35.5\% \Rightarrow 0.355$ $\quad$ *Move decimal two places to the left.*

Converting Percent to Fraction $\quad$ $15.5\% \Rightarrow 15\frac{1}{2} \div 100 \Rightarrow \dfrac{31}{2} \times \dfrac{1}{100} \Rightarrow \dfrac{31}{200}$

Converting Decimal to Percent $\quad$ $0.023 \quad \Rightarrow 2.3\%$ $\quad$ *Move decimal two places to the right.*

Converting Decimal to Fraction $\quad$ $0.625 \quad \Rightarrow \dfrac{625}{1000} \Rightarrow \dfrac{5}{8}$ $\quad$ *Rewrite as fraction, reduce.*

Converting Fraction to Decimal $\quad$ $\dfrac{3}{5} \Rightarrow 3 \div 5 \Rightarrow 0.6$ $\quad$ *Divide the numerator by the denominator.*

Converting Fraction to Percent $\quad$ $\dfrac{2}{3} \Rightarrow 2 \div 3 \Rightarrow 0.\overline{6} \Rightarrow 66\,\overline{6}\% \text{ or } 66\frac{2}{3}\%$

Percent Formula $\quad$ Old $\pm$ (%) of the Old = New

% as a ratio or comparison $\quad$ part : whole $\quad$ or $\quad$ $\dfrac{part}{whole}$

Simple Interest Formula $\quad\quad$ $I = P \times R \times T$

Compound Interest
Accrue

APR

Billing Cycle

Daily Periodic Rate

Purchase Balance

Pie Graph

Bar Graph

Use the given information to create the graph requested. Use a ruler and a protractor to measure your angles for the pie graphs. Be expressive and creative with colors.

10. Use the given information to create a pie graph.

     "Total 1998 cash receipts from marketing of meat animals fell to $43.6 billion. Cattle and calves accounted for 77 percent of this total, hogs and pigs 22 percent, and sheep and lambs 1 percent. Production increased for cattle and calves and hogs and pigs, but declined for sheep and lambs." (*United States Department of Agriculture, National Agricultural Statistics Service*)

11. Use the following table to create bar Graphs. (*Taken from Cattle and Calves: Average Price, Value per Head, and Total Value by State and United States, Revised 1998.*)

State	Cattle Price per 100 lbs	Calf Price per 100 lbs	Value per Head	Total Value for State
AL	$57.60	$79.10	$450.00	$697,500,000.00
AK	$62.00	$79.00	$730.00	$8,760,000.00
AZ	$62.10	$83.40	$660.00	$541,200,000.00
AR	$53.90	$78.80	$500.00	$900,000,000.00
CA	$48.80	$68.10	$740.00	$3,626,000,000.00
CO	$65.20	$86.20	$640.00	$2,080,000,000.00
CT	$50.00	$40.00	$730.00	$49,640,000.00
DE	$58.90	$63.50	$700.00	$19,600,000.00

a. Make a bar graph comparing Alabama's, California's, and Delaware's price per head for cattle.

b. Make a bar graph showing the comparison between the calf price per 100 pounds, for Colorado, Connecticut, Delaware, and Arizona.

c. Make a bar graph comparing the Total Value for each of the eight states.

12. Use the table in problem 11 to answer the following questions.

    a. Approximate how many head of cattle are in Alaska? Alabama?

    b. What is the approximate price per pound for calf in Colorado?

    c. What is the average weight of one calf in Arizona?

    d. What is the average weight of one cow in Arizona?

13. "... of these, 556 registered dairy cows from all six dairy breeds were named to the Elite Cow List: 10 Ayrshire, 53 Brown Swiss, 22 Guernsey, 430 Holstein, 36 Jersey, 3 Milking Shorthorn, and 2 Red and White." Use a pie graph to illustrate these findings.

14. Ninety Iowa cows qualified for the list of genetically superior grade dairy cows. The USDA list for High Ranking Grade Cows in Iowa included 2 Ayrshire, 2 Brown Swiss, 1 Guernsey, 79 Holstein, 1 Jersey, and 5 Red and White. Make a pie graph to illustrate what percent of the genetically superior dairy cows each type represents.

15. Make a bar graph showing the comparison by month of the price farmers were paid for their milk in the year 1996. The numbers are in dollars per CWT or 100 weight. (see the explanation on the next page)

        Jan 13.6,   Feb 13.4,   Mar 13.5,   Apr 13.4,   May 12.8,   Jun 12.6,
        Jul 12.2,   Aug 12.4,   Sept 12.8,   Oct 13,   Nov 13.1,   Dec 12.8.

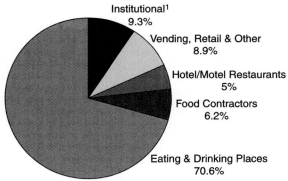

1997 Restaurant Industry Sales ($320 B)

Institutional[1]
9.3%

Vending, Retail & Other
8.9%

Hotel/Motel Restaurants
5%

Food Contractors
6.2%

Eating & Drinking Places
70.6%

[1]Businesses, Schools, Hospitals & Others with Own Food Services
(Source: *National Restaurant Association*)

16. Using the graph

   a. What percent of the Restaurant Industry Sales was the Institutional market?

   b. How many billion dollars were made from institutional sales?

   c. How many billion dollars were spent by Eating and Drinking Places?

   d. The Food Contractors spent how much in the Restaurant Industry?

   e. How much more was spent by Food Contractors than by the Hotel/Motel Restaurants? Give your answer in both percent and dollars.

Use the following graph for problems 17–25.

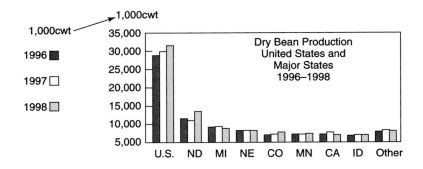

Hundredweight (Cwt)

A traditional unit of weight equal to 1/20 ton. The hundredweight is the English version of a commercial unit used throughout Europe and known in other countries as the quintal or the zentner. In general, this unit is larger than 100 pounds avoirdupois, so to fit the European market the hundredweight was defined in England as 112 pounds avoirdupois (about 50.8023 kilograms) rather than 100 pounds. This definition apparently dates from about the middle of the 1300's. In the United States, where there wasn't much need to align the unit with the quintal and zentner, the hundredweight came to *equal exactly 100 pounds* (about 45.3592 kilograms). To distinguish these two units, the British version is often called the long hundredweight and the American is called the short hundredweight or cental. The C in the symbol is of course, the Roman numeral representation for 100.

For problems 17–20, give your answers in Cwt and in pounds where appropriate.

17. Approximate how many pounds of dry beans were produced in Minnesota in 1997?

18. Approximate how many more pounds of dry beans were produced in North Dakota than in California in 1998?

19. What was the approximate total production of dry beans in pounds, in Michigan, for the years 1996–1998?

20. Approximately what percent of the total dry beans produced in the United States in 1996 were produced in Colorado?

21. Approximately how many pounds of dry beans were produced in Michigan in 1998? In the United States in 1997?

22. What percent of the total dry bean production in the United States in 1996 came from North Dakota?

23. Make a pie graph showing the dry bean production for the United States and major states in 1996.

24. In what year were the most beans produced? What state contributed the most?

25. Which state produced the most dry beans in 1997?

26. What percent of the dry beans came from "other states" in 1997?

27. Which states appear to be at a tie in wool production?

28. Which state produces the least amount of wool?

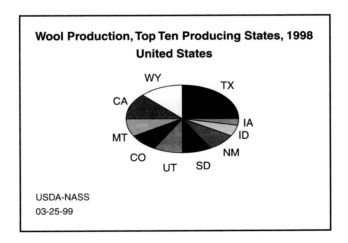

Wool Production, Top Ten Producing States, 1998
United States

WY
CA
TX
MT
IA
ID
CO
NM
UT
SD

USDA-NASS
03-25-99

29. What states are the top two producers of wool?

30. If you wanted to make a bar graph for the wool production shown in the pie graph above, what information would you put on the horizontal axis?

Vertical axis?

What information would you need to complete this bar graph?

31. For the following graph **All** represents the combination of breeding sheep and market sheep. **Brdg** stands for breeding only animals, and **Mkt** represents the market-bound sheep raised for food.

    a. How many thousand sheep are in Colorado?

    b. In Texas what percent of the sheep are for breeding and what percent are Market bound?

    c. Make a pie graph to show the distribution of sheep in the five states.

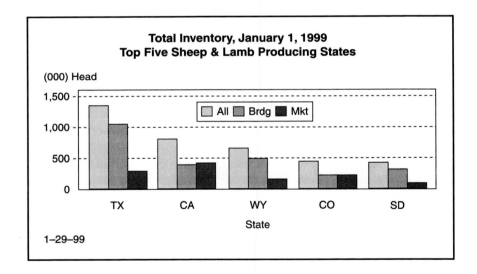

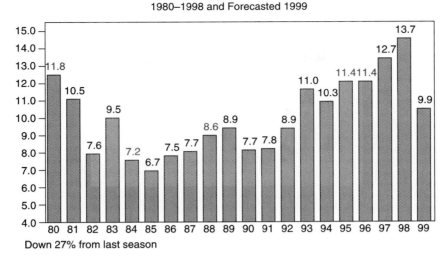

U.S. Orange Production
1980–1998 and Forecasted 1999

Down 27% from last season

32. In what year did the United States produce 9.5 million tons of oranges?

33. In what year(s) did the United States produce between 12 and 14 million tons of oranges?

34. In what year was the U.S. production of oranges the lowest?

35. How many more tons of oranges were produced in 1993 than in 1987?

Challenge Problem

36. What was the *average* number of oranges produced in the United States in the years 1980 and 1999?

37. Make a pie graph showing what percent of the total oranges produced each year represents. Use only 1995–1999.

38. Which consecutive five-year period produced the most oranges?

UNIT 3

Record Keeping

CHAPTER 8

Checking Accounts

Carrying a lot of cash is not a good idea for obvious reasons. It is possible to simply lose cash or, worst case, have it stolen. Another down side to carrying cash is that you usually don't remember where you have spent it. Have you ever wondered, "I had fifty bucks yesterday, now I have only ten. Where did it go?"

One way to manage your spending is by using a checking account. Often banks offer a debit card to accompany your checking account, which can double as a credit card. Well-organized checking accounts can be very helpful, but a mismanaged account can cause you an equal amount of grief. Here are some basic dos and don'ts for checking account use.

- *First Rule*: Never withdraw more than you deposit! This is often referred to as "bouncing a check" or "an overdrawn account." Most banks charge you a fee for overdrafts, and the business where you wrote the check does as well.

  Let's say you wrote a $65.75 check to Foot Locker® for shoes. Your checking account has only $45.50 left in it. What happens? First, Foot Locker deposits your check into its bank to collect the $65.75. Its bank gives it the money. Next, Foot Locker's bank takes your check to your bank to collect its money. Your bank sees that you do not have enough money in your account to cover the check and refuses to pay Foot Locker's bank. Your bank stamps the check "insufficient funds" and returns the check to Foot Locker's bank. Now, Foot Locker's bank calls the Foot Locker store and tells it to return the money that it has been given for your check. Sound complicated? It is.

  Finally, Foot Locker is stuck with a worthless piece of paper in return for the pair of shoes that you are now wearing. So, if you ever wondered why businesses charge for returned checks, now you know.

- *Second Rule*: KEEP GOOD RECORDS. Already discussed above is one disaster that could happen if you do not keep good records. What kind of records should you keep? A check register that is up to date and carried with your blank checks is the minimum. People with minimal computer skills can use a bookkeeping software such as Quicken® , MYM®—Manage Your Money, or one of many others that are on the market. These record keeping or accounting utilities are essential for any businesses and are an excellent tool for anyone wishing to keep track of how and where he or she spends their money.

- *Third Rule*: When you use your ATM (automatic teller machine) bank card, write down the amount you withdrew, along with any ATM fees, in your check register. Treat the total amount as if it was a check. Instead of recording a check number, use ATM. Don't forget to include any ATM fees.

- *Fourth Rule:* Learn to reconcile your checkbook with your bank statement. This is one way to check for accuracy of your records as well as the bank's records. Banks do make mistakes. Keep in mind that it is people like yourself who key in the check and deposit amounts to your account. And, although the banks have cross checks for accuracy, people do make mistakes.

- **Fifth Rule:** *Before you write a check, record it in your register.* This is the best advice I can pass on to you about checking accounts. When you are at Safeway® and the cashier tells you the total, record it in your check register *before* you write the check. Imagine looking in your check register only to notice that the last check you recorded was 456 when you are about to write check 458! You ask yourself, "What happened to 457?" By recording in your register *first*, you'll always know where you spent your money.

Notice in the register on the next page how the deposits were listed as **DEP** and the automatic teller machine withdrawals was listed as **ATM**. Some people like to keep a running total of their assets. It is a good idea and fairly easy if you carry a good calculator with you all the time. Just *keep good records*. That way, when you do sit down to figure your current balance, it will only be a matter of adding and subtracting.

```
┌─────────────────────────────────────────────────────────────────┐
│  Devin Thomas                                              221    │
│  321 Hwy F                                                        │
│  Ephraim, WI  54201                                               │
│                                          Date _____       │
│   Pay to the                                                      │
│   Order of  _____  $[      ]   │
│                                                                   │
│   _____  Dollars   │
│                                                                   │
│  Food Service Bank, Maui, Hawaii                                  │
│                                                                   │
│   Memo  _____   _____            │
└─────────────────────────────────────────────────────────────────┘
```

WRITING A CHECK

- Before writing a check, enter the transaction into your check register! This will insure that you never forget to whom you wrote a check.

- Always use a pen.

- Fill in the complete date, including the year: 5/6/99 is acceptable, although May 6, 2001, allows for fewer misinterpretations of your handwriting.

- On the first line—*Pay to the Order of*—write the name of the person or business you are paying.

- In the box to the right—$⬚—write the dollar and cents amount using a decimal point. For example, $3.45 or $47.68 or $115.25 or $1187.59.

- On the second line, write the dollar amount in words using the word **and** as your decimal point, followed by the numerical cent amount divided by 100. Then draw a line extending to the word *Dollars*.

        Forty-seven and 68/100 _____   *Dollars.*

- On the *memo* line, write a note to remind yourself what you bought.

- Sign your check with a unique signature that you will recognize always.

                                                    Your Full Name

THE CHECK REGISTER

Keeping track of your deposits, your bank machine withdrawals, and your checks is extremely important. Keep a neat, organized register, and you will make fewer mistakes. The following is an example of a check register.

CHECK #	DATE	CHECK ISSUED TO	PAYMENT		✓	DEPOSIT		BALANCE 875.00	
101	9/6	Class Act Restaurant	28	85				846	15
		Lunch with Renee							
DEP	9/18	gig at Hyatt				150	00	996	15
		Music income							
102	9/7	Costco	132	75				863	40
		groceries							
ATM	9/28	Cash for movie	40	00				823	40
		Recreation							

THE DEPOSIT SLIP

A deposit slip is fairly easy to fill out. *Currency* is the name given to paper money; any denomination of bills. *Coins* of course means just that. *Check*s have two columns per check. For a personal account you may write yourself a memo, such as who gave you the check and for what reason, in the first column. However, for a business account, it is customary to write the bank* number that is on the check you are depositing in the first column. The second column is reserved for the check amount.

Suppose that you are depositing checks into your account and decide that you need a little cash. First you enter the amount of cash that you want on the *Less Cash Received* line. It is then subtracted from the

Maui Community College Bank
Checking Deposit

Currency		
Coins		
	Salary	865.00
C	Gig	150.00
h e	Mom	250.00
c k		
s		
Total		1265.00
Less Cash Receives		60.00
Net Deposit		1205.00

Date _____ 6/8/02 _____

Print Name _____ Nathan Spring _____

Sign for Cash _____ Nathan Spring _____

Account 12345-67890

*Each bank has a unique identification number that is printed on all checks. It is usually located next to the printed check number in the upper right hand corner. These numbers, which are unique to each bank, code the bank name, which branch, and the state in which the account resides. A bank identification number is also printed to the left of your personal account number, along the bottom edge of each check and deposit slip. Here is an example of a bank code number: $\frac{37-501}{1213}$.

sum total of the checks, coins, and currency you are depositing, giving you the *Net Deposit* amount. The bank teller will then ask you to put your signature directly under your printed name where it says *Sign for cash*. This final signature is to acknowledge that you were given the cash at the time of deposit.

BUSINESS CHECKING AND CATEGORIZING

If you open a small business, you will need a business checking account. You will need to track your expenditures very closely for income tax purposes. Keep in mind that certain business expenditures are totally deductible from your business income, while for other items, the government may allow only a percent of the expenses to be deducted. It is necessary therefore to set up good record-keeping habits from the start.

First of all, you may want to set up categories for your spending habits. Make a note in your register which category each check fits into. For instance you may have a category titled "equipment repair," which could be a subcategory of "fixed assets."

A useful main category for your personal account may be "medical expenses." This category could include subcategories of doctor, dentist, eye care, and pharmacy expenses. There are many ways to categorize expenses, so don't worry if it takes a while to create a system that works for you.

For now, practice categorizing with your personal checking account. Try some basic categories like, "food," "clothing," "recreation," and "rent." You may be surprised by what you find at the end of the year when you total each category.

THE BANK STATEMENT

Once each month, your bank sends you a statement. The statement contains information on:

1. Your opening balance

2. A summary of your deposits

```
Medical
   Dental
   Doctor
   Mental Health
   Prescription Drugs
   Vision Care Plan
```

```
Pets
   Food
   Vet
```

```
Utilities
   Cable TV
   Gas & Electric
   Telephone
   Trash Collection
   Water
```

```
Business Deductions
   Postage & Delivery
   Expenses
   Equipment
   Repairs
   Supplies
   Office Expenses
   Computer
   Printing/Copies
```

```
Vacation
   Lodging
   Motel/Hotel Costs
   Transportation
   Recreation
```

Just a few basic category and subcategory suggestions.

3. A list of the checks that you have written that have been paid by your bank and their amounts

4. Any service charge that the bank charges and has been withdrawn from your account

5. The dates and amounts of your ATM transactions

6. Your ending balance

page 1 of 2

WATSON BANK
Member FDIC

STATEMENT OF ACCOUNT
Account Number 1234-567
Opening Balance $3,641.05
Statement Date 4/26/01–5/20/01

Date	Check #	Amount	Date	Check #	Date	ATM Transaction	Daily Balance
4/26/01	427	137.67					3,503,38
5/10/01	433 *	63.80	5/1/01	1578.00	5/6/01	60.00	4,957.58
5/10/01	434	20.00					4,937.58
5/11/01	435	200.00					4,737.58
5/12/01	436 *	156.24			5/12/01	40.00	4,541.34
5/12/01	446	62.40					4,478.94
5/12/01	447	23.09			5/12/01	20.00	4,435.85
5/13/01	448	500.00					3,935.85
5/14/01	449	25.00					3,910.85
5/14/01	451 *	104.35	5/15/01	2345.88			6,152.38
5/17/01	452	20.00			5/16/01	100.00	6,032.38
5/17/01	453	70.05					5,962.33
5/18/01	454	12.00					5,950.33
5/18/01	455	200.00			5/18/01	60.00	5,690.33
5/20/01	456	257.88					5,432.45

Checking Account Summary

Opening Balance on 4/26	$ 3,641.05
Total of 2 Deposits for	$ 3,923.88
Total of 15 Checks for	$ 1,852.48
Total of 5 Other Debits	$ 280.00
Closing Balance on 5/20/01	$ 5,432.45

Monthly statement of account.

To understand the Statement of Account, let's begin by looking at the bank statement *OPENING BALANCE*. This amount can also be found on the previous statement labeled *Closing Balance on 4/25/01*. The balance amount $3,641.05 is entered in the upper right margin beneath the *ACCOUNT NUMBER*.

On the first line, date 4/26/01, Check 427 has been deducted from the opening balance, and the new *DAILY BALANCE* is the result of this subtraction.

line 1 $3,641.05 – $137.67 = $3,503.38

On the second line, date 5/10/01, there is an asterisk * next to check 433. This asterisk signifies that there is one or more checks in the sequence that are missing. In this case you will notice that checks numbered 428 – 432 are not listed. These checks may not have cleared the bank yet, or they may have been listed on the previous statement, in which case they would have already been checked off in the check register.

On this same line, you see that on 5/1/01 $1,578 has been deposited into the account, and on 5/6/01 the automatic bank teller card was used to withdraw $60. These transactions brought the balance to $4,957.58.

$$\text{line 2} \quad \$3,503.38 - \$63.80 + \$1,578 - \$60 = \$4,957.58$$

daily balance   check #433   deposit   ATM

The *ACCOUNT SUMMARY* reflects the whole month's transactions. It is here that you can cross check the register as to the number of checks that have cleared, the total number of deposits the bank has received, and of course the opening and closing balance for the month. Note the line that lists "*Total of 5 Other Debits.*" The word *debit* here means an amount that has been subtracted from the account. In this case it refers to the five automatic teller machine withdrawals for the month.

> It is worth noting here that not all bank statements use the same format. There are many types of statements, some easier to read than others, all of which show completed transactions for the month listed.

RECONCILING YOUR ACCOUNT

It is imperative that you reconcile your checkbook register with your bank statement each month. However, if you skip a month or two, don't panic or give up. Although it may take longer, dig in the minute you have time. The longer you wait, the harder it is to protest any bank errors you may find, not to mention your own errors, which could cause you to bounce checks.

The first step to reconciling your account is to check (✓) off all check numbers that are listed on your current bank statement. *This is done on your check register.* If you look back to the sample check register you will note the ✓ column. This column allows you to place a mark next to each check that has cleared the bank.

On the back of most bank statements you will find a form that walks you through the reconciling process. Here is an example of the previous statement being reconciled (see top of next page).

STEP 1

List all checks that you have written, along with any ATM withdrawals that you have made, that have not yet cleared or been turned into the bank for collection. The checks that have not cleared the bank are the ones not checked off in your check register, as described in the previous two paragraphs.

Also mentioned previously, on many statements there is an asterisk * to indicate where a check is missing from the numerical sequence. This makes it easy to see which checks have cleared the bank and which ones are still outstanding.

For this statement, checks numbered 428–432, 437–445, and 450 have not cleared the bank. To complete STEP 1 you need to write down each check and its corresponding amount in the space provided.

STEP 1

OUTSTANDING DEBITS

List all debits that have not been reported on this or prior statements

CHECK NO.	DEBIT AMOUNT
428	16.78
429	115.25
430	300.00
431	15.00
432	9.00
***	***.**
437	11.15
438	28.98
439	146.78
440	687.00
441	89.25
SUBTOTAL	1419.19

CHECK NO.	DEBIT AMOUNT
442	88.09
443	50.00
444	25.50
445	67.48
***	***.**
450	250.00
SUBTOTAL	1419.19
TOTAL	1900.26

STEP 2

CLOSING BALANCE on this statement $ 5432.45

Add deposits that
are not listed on
this statement

$ 1125.50
     280.39
     24.75

Total Deposits   1430.64   + $ 1430.64

Subtotal = $ 6863.09

Deduct outstanding debits...   – $ 1900.26

TOTAL   This amount should equal your checkbook balance   = $ 4962.83

Next, add the first column of check amounts labeled *DEBIT AMOUNTS* and put the total on the line labeled SUBTOTAL. Write this amount in the *second column as well*, next to the word subtotal.

Now add the debit amounts in column 2, along with the subtotal from column 1 to find a *TOTAL* of your outstanding debits. This may seem a bit awkward, but in trying to save space, the bank has made two short ledgers where one longer one would have been more appropriate.

STEP 2

a. Begin with the *CLOSING BALANCE* written on your statement, $5,432.45.

b. *ADD ANY DEPOSITS NOT SHOWN ON YOUR STATEMENT.* You have made three deposits that do not appear on your statement. List these amounts: $1,125.50, $280.39, and $24.75.

c. Find the *total* of the three *DEPOSITS.* $1,430.64.

d. Add the statement *closing balance* quantity to the *OUTSTANDING DEPOSITS* total for a subtotal of $6,863.09.

e. Subtract or **Deduct the outstanding debits amount**, the total from Step 1, of $1,900.26 from the *SUBTOTAL.*

f. This *total,* $4,962.83, should be the exact balance written on the last line in your checkbook register. It is **your current actual balance**.

Keep in mind that the "current balance" shown on an ATM slip can fool you! It does *NOT* reflect your actual current balance. As you have just seen, there may be outstanding checks not figured into the balance given. BE CAREFUL! Always know your real balance.

Exercise Set 8

1. To write a check, you have to write out the numbers in words. Spell the following numbers correctly. (*Hint: Use a hyphen in numbers like twenty-two.*)

   a. 11   b. 40   c. 90   d. 88   e. 14   f. 49

2. Fill out the deposit slip:
   a. $46.00 in currency, $0.87 in coin
   b. Paycheck in the amount of $650.87
   c. Date: May 16, 2001

Maui Community College Bank Checking Deposit			Currency		
			Coins		
			C		
			h		
Date			e		
			c		
			k		
Print Name			s		
			Total		
Sign for Cash			Less Cash Receives		
			Net Deposit		
Account 12345-67890					

3. a. $ 12.39 in coins
   b. Check – Income tax refund for $655.48
   c. Check – Gig at the Embassy Suits $225.00
   d. Check – From cookies you sold $38.78
   e. Date: October 1, 1999

Maui Community College Bank Checking Deposit			Currency		
			Coins		
			C		
			h		
Date			e		
			c		
			k		
Print Name			s		
			Total		
Sign for Cash			Less Cash Receives		
			Net Deposit		
Account 12345-67890					

4. a. Check – Car you sold $1,800.00
   b. Check – Birthday check $250.00
   c. Check – Sold your books $115.00
   d. Less cash – $300.00
   e. Date: December 31, 1999

Maui Community College Bank Checking Deposit		Currency		
		Coins		
		C		
		h		
Date		e		
		c		
		k		
Print Name		s		
		Total		
Sign for Cash		Less Cash Receives		
		Net Deposit		
Account 12345-67890				

5. a. Check – Jean loan payback $20.00
   b. Paycheck – $360.35
   c. Check – Fernando, chair $76.00
   d. Less Cash – $100.00
   e. Date: April 5, 2002

Maui Community College Bank Checking Deposit		Currency		
		Coins		
		C		
		h		
Date		e		
		c		
		k		
Print Name		s		
		Total		
Sign for Cash		Less Cash Receives		
		Net Deposit		
Account 12345-67890				

Fill out a check for each problem below. Be sure to make up a memo for each purchase, watch your spelling and remember the dashes in some double digit numbers.

6. On 4/5/89 you spent $28.35 at Long's Drugs.

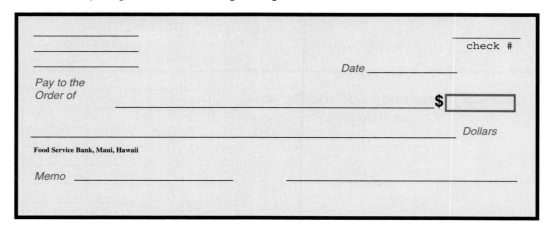

7. G.T.E. on June 3, 1999 for $63.45.

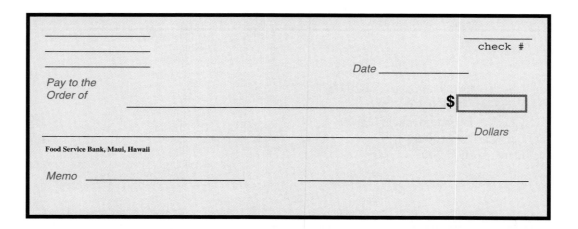

8. Aloha Airlines to Kauai, 5/6/01, $75.8.

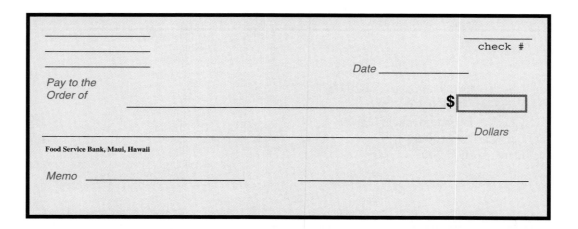

9. Borders Bookstore® , $38.46, September 16, 2002.

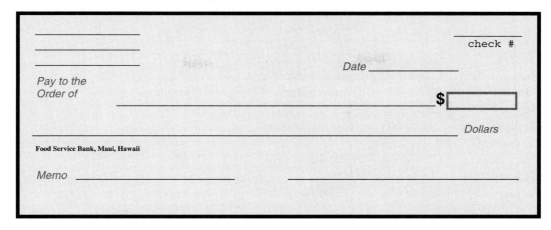

10. Maui Dive Shop, $268, 12/3/03.

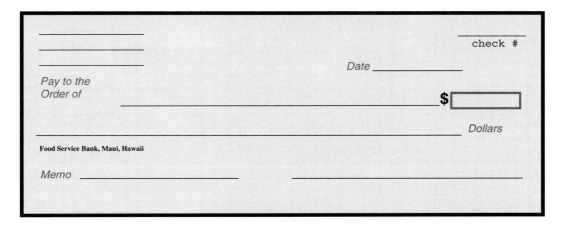

11. Eagle Discount Store, plants, $49.96 on 6/1/2004.

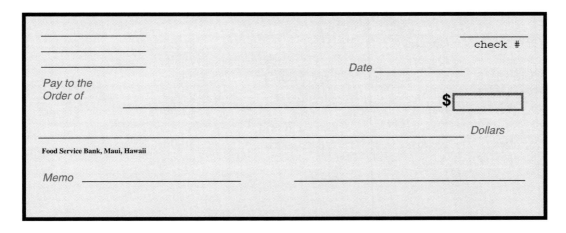

Fill in a check register using the given information. Use dates without a year i.e., 12/4. Draw a current balance on each line of the register. Use the memo as needed.

12. Beginning balance on January 1, 2003 $238.75

January 7	Check 601	Walden's Books	$ 22.37
January 9	Check 602	Cyds Motorcycles	$ 85.26
January 9	ATM	Cash Machine	$ 60.00
January 11	Deposit		$1,735.95
January 12	Check 603	Sports Authority	$ 33.38
January 15	ATM		$ 20.00
January 18	Check 604	Foodland	$ 19.46
January 20	Deposit		$500.00
January 21	Check 605	Class Act	$ 18.50

CHECK#	DATE	CHECK ISSUED TO	PAYMENT	✓	DEPOSIT	BALANCE	

13. Balance forward on October 3, 2002 $615.00

October 3	Check 245	Pet Shop	$ 46.55
October 4	Deposit		$509.00
October 5	Check 246	College Book Store	$ 54.66
October 5	ATM		$100.00
October 8	Deposit		$382.00
October 13	Check 247	Waterbed World	$667.83
October 17	Deposit	Birthday Checks	$450.00
October 18	ATM		$ 80.00

CHECK#	DATE	CHECK ISSUED TO	PAYMENT	✓	DEPOSIT	BALANCE

14. Your balance forward July 2, 2004 $128.90

July	2	Check 505	Sam's Fireworks	$107.88
July	3	Check 506	Coffee Works	$14.50
July	3	ATM	refused *sorry no money available*	
July	6	Deposit		$148.36
July	19	Check 507	Kmart	$22.05
July	20	Check 508	IHOP	$18.50
July	26	Check 509	Christ Church	$20.00
July	28	Check 510	Safeway	$56.78

CHECK#	DATE	CHECK ISSUED TO	PAYMENT	✓	DEPOSIT	BALANCE

15. Your balance forward on December 12, 2007 $5,349.25

December 12	Check 1006	Toyota	$450.55
December 13	Check 1007	Christmas Shop	$46.79
December 14	ATM	$160	
December 15	Deposit	paycheck	$2,564.77
December 16	Check 1007	St. Anthony School	$1,056
December 18	Check 1008	Longs Drugs	$ 36.21
December 19	Check 1009	Dr. Baum	$88.45
December 20	Check 1010	Toys R Us	$489.99
December 23	ATM	$200.00	

CHECK#	DATE	CHECK ISSUED TO	PAYMENT	✓	DEPOSIT	BALANCE	

For problems 16 and 17 use the forms provided on the next page.

Here is a portion of a check register and a bank statement. After comparing this register with its accompanying bank statement, reconcile the register. Use the given reconcile forms.

a. Calculate, then enter the actual current balance in the register.

b. Put a ✓ in the appropriate column on the register next to the check numbers that appear on the statement.

c. On the reconcile sheet, list each check and ATM withdrawal that does not appear on your statement. Add them up.

d. Complete the daily balance column for the bank statement.

e. List any deposits that do not appear on your statement. Complete the reconcile process.

f. Does your check register balance match your reconciled balance?

For use with 16.

STEP 1

OUTSTANDING DEBITS

List all debits that have not been reported on this or prior statements

CHECK NO.	DEBIT AMOUNT		CHECK NO.	DEBIT AMOUNT
			SUBTOTAL	
SUBTOTAL			TOTAL	

STEP 2

CLOSING BALANCE on this statement $ _____

$ _____

_____

Add deposits that   _____

are not listed on   _____

this statement   _____

_____ + $ _____

Subtotal = $ _____

→ Deduct outstanding debits. . . − $ _____

TOTAL   This amount should equal your
checkbook balance   = $ _____

For use with 17.

STEP 1

OUTSTANDING DEBITS

List all debits that have not been reported on this or prior statements

CHECK NO.	DEBIT AMOUNT		CHECK NO.	DEBIT AMOUNT
			SUBTOTAL	
SUBTOTAL			TOTAL	

STEP 2

CLOSING BALANCE on this statement $ _____

$ _____

_____

Add deposits that   _____

are not listed on   _____

this statement   _____

_____ + $ _____

Subtotal = $ _____

→ Deduct outstanding debits. . . − $ _____

TOTAL   This amount should equal your
checkbook balance   = $ _____

16. Opening balance: $388.40.

	WATSON BANK		STATEMENT OF ACCOUNT				

WATSON BANK
Member FDIC

STATEMENT OF ACCOUNT
Account Number 1234-567
Opening Balance $368.40
Statement Date 8/501–8/21/01

Date	Check #	Amount	Date	Check #	Date	ATM Transaction	Daily Balance
8/9/01	512	12.56					
8/18/01	514*	28.09	8/13/01	150.38	8/10/01	40.00	
8/18/01	515	33.44					
8/21/01	516	89.67					

Checking Account Summary

Opening Balance on 8/5/01	$	388.40
Total of 1 Deposits for	$	150.38
Total of 4 Checks for	$	163.76
Total of 1 ATM Transaction	$	40.00
Closing Balance on 5/20/01	$	335.02

$388.40

512	8/7	Door County Advocate	12	56						
		magazine								
513	8/9	Skyway Drive In	15	78						
		lunch								
ATM	8/10	Cash	40	00						
Dep	8/13	Deposit			150	38				
514	8/14	Johnny Seed Flowers	28	09						
		ftd to mom								
515	8/15	Dr Anzi	33	44						
		glasses								
ATM	8/15	Cash	20	00						
516	8/17	Piggly Wiggly	89	67						
517	8/18	Dunkin Donuts	7	12						
		Dozen for work								

17. Opening Balance $1,479.39.

											$1,479.39	
646	3/7	Las Cruces News	28	75								
		paper										
647	3/9	Costco	239	48								
		tool box										
648	3/10	Dells Farm Market	36	77								
		dog food										
Dep	3/13	Deposit					560	38				
		pay check										
649	3/14	Cut-Rate Tickets	460	59								
		Vegas trip										
ATM	3/16	Cash	20	00								
650	3/17	Iditarod Trail	100	00								
		donation										
Dep	3/22	Deposite					125	00				
ATM	3/23	Cash	80	00								

WATSON BANK
Member FDIC

STATEMENT OF ACCOUNT
Account Number   5555-121212
Opening Balance $1,479.39
Statement Date   3/7/03–3/21/03

Date	Check #	Amount	Date	Check #	Date	ATM Transaction	Daily Balance
3/11/03	646	28.75					
3/12/03	647	239.48	3/13/03	560.38			
3/18/03	649*	460.59			3/16/03	20.00	
3/21/03	650	100.00					

Checking Account Summary		
	Opening Balance on 3/7/03	$ 1,479.39
	Total of 1 Deposits for	$ 560.38
	Total of 4 Checks for	$ 828.82
	Total of 1 ATM Transaction	$ 20.00
	Closing Balance on 3/21/03	$ 1,190.95

CHAPTER 9

Price Lists, Requisitions, Purchase Orders, and Invoices

PRICE LISTS

Many wholesale distributors of food products change their prices so often that they no longer print a bound catalog with their price lists. Instead, some sales representatives carry laptop computers with daily prices that they will printout for you by request. Some companies will fax computer printout price lists on the day that you intend to order. Still others will only quote prices over the phone and make sure you know that they are "subject to change." Many will print out a hard copy product list, but without a price column.

ITEM	DC Coffee & Tea	Sam's Wholesale	Sysco Inc	EconoFoods	Karps Bakers	Cedar Crest Ice Cream	Orchard Country	DATE	Vendor Phone Number
Sweetie Pies — Vendor Price Comparison Worksheet									
Milk Evaporated									984-3948
Pumpkin Libby's		$2.90	$3.40	$3.10			$2.69	7/5/02	984-0394
Colombian Supremo									875-9584
Sarsaparilla Syrup									839-6972
Cherries GLZD									839-4811
Fruitcake Batter									244-3773
Puff Paste Bulk Eg									839-2546
Pie Foil Pan 9"									839-3399
Pie Foil Pan 10"									868-0007
Vanilla Ice Cream									555-2434

Some of the small business owners that I have interviewed create their own spreadsheets to use for comparison of prices. The products they frequently purchase are listed in the leftmost column followed by a blank column for each vendor from which they purchase supplies.

When they phone the distributor for updated prices, they pencil the current price under the distributor's name. In this manner they are able to compare their distributors' prices for each item, at a glance, and make a wise purchase choice.

REQUISITIONS

Requisitions are used to request supplies needed for a particular operation. If you work for a large establishment, such as a four-star hotel, there is most likely a stockroom or warehouse on the premises in which supplies needed to keep the hotel running are kept.

Let us assume for a moment that you are the salad prep for the pool bar, and you need tomatoes, lettuce, and celery for some salads on the lunch menu. You will fill out a *requisition* form to let the management know what you need. Someone assigned to fill your requisition will then retrieve the salad supplies.

Depending on the way your hotel works, you will probably only fill out the amount requested. The other columns will be filled in by the stockroom manager and sent back to you with your supplies. This way you will be aware of the cost of each item you use, which will in turn make it possible for you to figure the cost of each dish you prepare *(discussed in Unit 4)*.

SWANK HOTEL *Pool Dining*

Requisition for Supplies

QTY	PK	SIZE	DESCRIPTION	CODE	COST	PER	YIELD	UNIT PRICE	EXTENDED AMOUNT
2	BG	25LB	TOMATOES						
1	CS	36 STLK	CELERY						
3	CS	24HD	LETTUCE ICEBERG						
1	BG	50LB	ONIONS						
1	CS	14/1#BG	RADISHES CELLO						

Requisition #1

Example 1: Find the total amount of each food ordered.

- 2 bags of tomatoes, each 25 pounds   50 pounds of tomatoes

- 1 case of celery, 36 stalks each case   36 stalks of celery

- 3 cases of iceberg lettuce, 24 head per case   72 head of lettuce

- 1 bag of onions, 50 pounds per bag   50 pounds of onions

- 1 case of radishes, 14 1-pound bags per case   14 pounds of radishes

PURCHASE ORDERS

Let's assume you own your own restaurant and your cook requests supplies. Depending on the size of your restaurant, you may have your own requisition forms that your cook has filled out informing you of each item he or she needs. If not, possibly your cook has scribbled it on a note pad or perhaps told you verbally what was needed.

At any rate, you now have some kind of list indicating the items you need to purchase from your suppliers. You may choose to:

1. Fill out a purchase order and fax it to the vendor.

2. Phone in your order directly to a supplier.

3. Get into your car and go purchase the necessary items yourself.

The following is an example of a *purchase order* that was designed for you to fill in and fax. Notice that the prices are not shown, and that the form is made exclusively for the business, Arts Inn So's Esther. The vendor is PKJ Foodservice.

PKJ Foodservice
345 Forest Ave.
Kihei, HI

CUSTOMER ORDER GUIDE
ORDER GUIDE #13245

Arts Inn So's Esther
243 Apau Place, Makawao

CONFIDENTIAL

3/4/05

QTY	PK	SIZE	DESCRIPTION	CODE	COST	PER	YIELD	UNIT PRICE	EXTENDED AMOUNT
2	4	GAL	DRESSING BUTTERMILK	890764		OZ	512		
1	18	GAL	DRESSING MIX BLEU CHEESE	801309		OZ	2304		
5	88	CT	APPLE DELICIOUS GOLDEN	60905P		PC	88		
3	6	#10	APPLE SAUCE MICHIGAN	603942		OZ	648		

PKJ Foodservice

Requisition #2

You enter the *quantity only*. When it is received, the distributor enters the current unit price, cost per yield, the extended amount and the totals. As you read the purchase order form, notice the *use* of the *PK* and *SIZE* column is different than in Requisition #1. Other companies' purchase order forms may look quite different than either of these two.

Example 2: On the previous purchase order form, look at the third line down, the Golden Delicious APPLES. You have ordered 5 crates of apples. Each crate holds 88 apples.

    a.   Find the cost per apple.

$$\text{Total Cost} \div \# \text{ of Units} = \text{Cost per Unit}$$

$$\$29.88 \div 88 = \$0.3395 \Rightarrow \cong \$0.34 \text{ per apple}$$

    b.   Find the extended or total amount for the apples ordered if the *unit* price (price per crate) is $29.88.

$$\text{Cost Per Unit} \times \# \text{ of Units ordered} = \text{Extended Amount}$$

$$\$29.88 \times 5 = \$149.40$$

Worth a Note

In the previous requisition example, the *PK* column stood for packaging, while the *SIZE* column told you how many and how big each package was. On this purchase order, the *PK* column tells you *how many* are in a package, while the *SIZE* column tells the *type of package*.

Now that you have seen examples of different types of price lists, purchase orders, and requisitions, let's get into the mathematics of invoices. Keep in mind that understanding the mathematics of an invoice is one of the necessary tools needed for pricing your menu, which is an important key to making a profit.

INVOICES

An invoice is an accounting of how much you spend with a particular supplier on a given order. As you will see in the section on "Income Statements," these invoices are what make up the "cost of goods sold." Let's look at a sample invoice on the next page.

In these forms some standard abbreviations are used to indicate the types of containers used for packaging.

- CS - case
- SAK - sack
- BG - bag
- CNT - count
- BX - box
- CTN - carton
- EA - each
- PC - piece
- CT - crate

For example, a lot of information is contained on each line of an invoice. Examine the first line of the invoice from **Reinhart FoodService** ® (see top of next page).

   5 CS 35/LBS RHHP 15100 OIL:CLEAR VEG FRYING 560 1 OZ .020 11.37 56.85

Five cases, *5 CS*, of clear vegetable frying oil, brand *RHHP*, **Reinhart FoodService** item number 15100 have been shipped.

The *unit* is a case, which contains 35/LBS, or 35 one pound containers. That would make a total of 35 × 5 = 175 pounds of oil shipped.

Next, look at the *portion* section. The yield is listed as the number of recipe units. The yield or *# of ru* is 560. 16 ounces per pound × 35 pounds of oil = 560 ounces of oil.

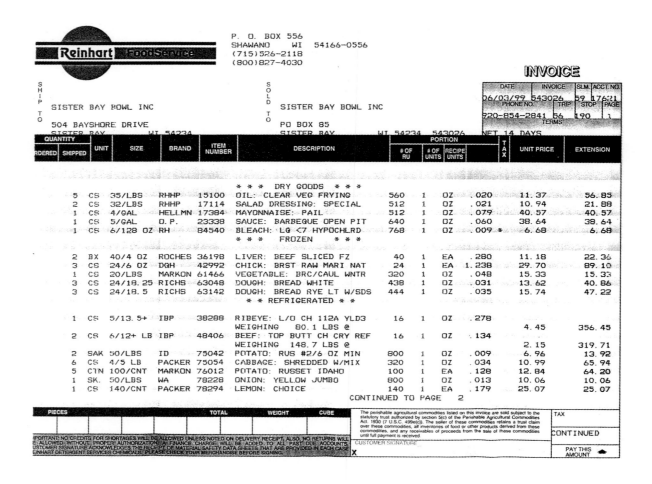

Reinhart FoodService

P. O. BOX 556
SHAWANO   WI   54166-0556
(715)526-2118
(800)827-4030

INVOICE

DATE	INVOICE	SLM.	ACCT. NO.
06/03/99	543026	59	17621

PHONE NO.	TRIP	STOP	PAGE
920-854-2841	56	190	1

TERMS
NET 14 DAYS

SHIP TO:
SISTER BAY BOWL INC
504 BAYSHORE DRIVE
SISTER BAY   WI 54234

SOLD TO:
SISTER BAY BOWL INC
PO BOX 85
SISTER BAY,   WI 54234   543026

QUANTITY ORDERED	SHIPPED	UNIT	SIZE	BRAND	ITEM NUMBER	DESCRIPTION	# OF RU	# OF UNITS	RECIPE UNITS	TAX	UNIT PRICE	EXTENSION
						*** DRY GOODS ***						
5		CS	35/LBS	RHHP	15100	OIL: CLEAR VEG FRYING	560	1	OZ	.020	11.37	56.85
2		CS	32/LBS	RHHP	17114	SALAD DRESSING: SPECIAL	512	1	OZ	.021	10.94	21.88
1		CS	4/GAL	HELLMN	17384	MAYONNAISE: PAIL	512	1	OZ	.079	40.57	40.57
1		CS	5/GAL	O.P.	23338	SAUCE: BARBEQUE OPEN PIT	640	1	OZ	.060	38.64	38.64
1		CS	6/128 OZ	RH	84540	BLEACH: LG <7 HYPOCHLRD	768	1	OZ	.009 *	6.68	6.68
						*** FROZEN ***						
2		BX	40/4 OZ	ROCHES	36198	LIVER: BEEF SLICED FZ	40	1	EA	.280	11.18	22.36
3		CS	24/6 OZ	DQH	42992	CHICK: BRST RAW MARI NAT	24	1	EA	1.238	29.70	89.10
1		CS	20/LBS	MARKON	61466	VEGETABLE: BRC/CAUL WNTR	320	1	OZ	.048	15.33	15.33
3		CS	24/18.25	RICHS	63048	DOUGH: BREAD WHITE	438	1	OZ	.031	13.62	40.86
3		CS	24/18.5	RICHS	63142	DOUGH: BREAD RYE LT W/SDS	444	1	OZ	.035	15.74	47.22
						** REFRIGERATED **						
1		CS	5/13.5+	IBP	38288	RIBEYE: L/O CH 112A YLD3 WEIGHING 80.1 LBS @	16	1	OZ	.278	4.45	356.45
2		CS	6/12+ LB	IBP	48406	BEEF: TOP BUTT CH CRY REF WEIGHING 148.7 LBS @	16	1	OZ	.134	2.15	319.71
2		SAK	50/LBS	ID	75042	POTATO: RUS #2/6 OZ MIN	800	1	OZ	.009	6.96	13.92
6		CS	4/5 LB	PACKER	75054	CABBAGE: SHREDDED W/MIX	320	1	OZ	.034	10.99	65.94
5		CTN	100/CNT	MARKON	76012	POTATO: RUSSET IDAHO	100	1	EA	.128	12.84	64.20
1		SK.	50/LBS	WA	78228	ONION: YELLOW JUMBO	800	1	OZ	.013	10.06	10.06
1		CS	140/CNT	PACKER	78294	LEMON: CHOICE	140	1	EA	.179	25.07	25.07

CONTINUED TO PAGE   2

PIECES	TOTAL	WEIGHT	CUBE		TAX

The perishable agricultural commodities listed on this invoice are sold subject to the statutory trust authorized by section 5(c) of the Perishable Agricultural Commodities Act. 1930 (7 U.S.C. 499e(c)). The seller of these commodities retains a trust claim over these commodities, all inventories of food or other products derived from these commodities, and any receivables of proceeds from the sale of these commodities until full payment is received.

CONTINUED

CUSTOMER SIGNATURE
X

PAY THIS AMOUNT

IMPORTANT: NO CREDITS FOR SHORTAGES WILL BE ALLOWED UNLESS NOTED ON DELIVERY RECEIPT. ALSO, NO RETURNS WILL BE ALLOWED WITHOUT PROPER AUTHORIZATION. A FINANCE CHARGE WILL BE ADDED TO ALL PAST DUE ACCOUNTS. CUSTOMER SIGNATURE ACKNOWLEDGES THE RECEIPT OF MATERIAL SAFETY DATA SHEETS THAT ARE PROVIDED IN EACH CASE REINHART DETERGENT SERVICES CHEMICALS. PLEASE CHECK YOUR MERCHANDISE BEFORE SIGNING.

The size of each *RU* is one ounce. Shown *# OF UNITS|RECIPE UNITS, as 1 OZ*. We conclude that there are 560 one-ounce recipe units in *each case* of *oil:clear veg frying*.

Notice that the heading of the next column is *RU PRICE*, or the price per ounce in this case, which is *.020*. The *RU PRICE* is found by dividing the *unit price* by the total number of recipe units. As one unit contains 560 1-oz recipe units of oil, thus:

$$\Rightarrow \$11.37 \div 560 \quad \cong \quad \$0.0203 \text{ per ounce.}$$

Next to last is the *UNIT PRICE*. This is the price of *one case* of oil, or $11.37.

Last, the *EXTENSION* price. This is the price of one case extended to the cost of the five cases you ordered. Your cost for five cases $\Rightarrow \$11.37 \times 5 = \$56.85$.

Example 3:   Reading and interpreting the sample **Reinhart FoodService** ® invoice above:

a.   How much would one *POTATO: RUSSET IDAHO* cost?

One potato would cost $0.128, which is ≈ 13¢. Found under the *PORTION* section. Notice that the recipe unit is *each*.

b. How much would 5 ounces of *ONION: YELLOW JUMBO* cost?

The price is listed as $0.013 per ounce. To find the cost of five ounces, $\Rightarrow$ 5 x $0.013 = $0.065 which is $\approx$ 7¢ for 5 oz of onion.

c. Find the weight of one chicken breast, raw, marinated.

In the *SIZE* column on the *CHICK: BRST RAW MARI NAT* line, it says 24/6 oz. This means that there are twenty-four 6 oz breasts in one case. Therefore one chicken breast weighs 6 ounces.

d. How much does one ounce of the chicken breast cost? One pound?

Each chicken breast costs $1.238, as found in the *PORTION SECTION*. Each breast weighs 6 ounces as found in part c above. Now just divide the cost by the total ounces to find the cost per ounce: 1.238 ÷ 6 = 0.2063. One ounce of chicken breast costs approximately 21¢, while one pound is 16 times that amount, or approximately $3.30.

e. How many individual bags of shredded cabbage did you receive?

If you look under the *SIZE* column on the cabbage line, you see *4/5 LB*, which means that there are four 5 lb bags in each case. You ordered 6 cases, so you received $\Rightarrow$ 6 x 4 (bags) = 24 bags of shredded cabbage.

f. How many gallons of Open Pit Barbecue Sauce did you order?

Each case of BBQ sauce contains 5 one-gallon containers. Therefore, you ordered 5 gallons of BBQ SAUCE.

g. How was the RU price found for the *BBQ SAUCE*?

The RU price was found by dividing the unit price by the number of units.
$38.64 per case ÷ 640 oz. per case

h. What is the package yield for the *LEMONS*?

The lemons come in a case, which yields 140 lemons.

This next illustration is of an invoice that *does not have a portion section*. On this type of invoice, you will need to do a lot of arithmetic to find out the answers to some very important cost questions.

Notice that the column headings are once again different than in the previous examples. The first column QTY is the quantity ordered, the CONT column tells the type of container or packaging, while the PK (pack) column tells how many of the units listed under the SIZE column are in the container. The first line reads 4 cases, each of which contain two 5-pound cans of peaches.

System Food Service Distributors
1 Food Plaza North, Jackson WY

QTY	CONT	PK	SIZE	DESCRIPTION	ITEM CODE	UNIT PRICE	EXTENDED AMOUNT
4	CS	2	5LB	SYS IMP PEACH SLICED	1024346	16.19	64.76
2	BG	1	50LB	C&H SUGAR GRANULATED	4566055	20.89	41.78
1	CS	36	1LB	PACKER BUTTER SOLID WIS	1160089	48.78	48.78
3	BG	1	50LBS	GOLDMDL FLOUR UNBLCH	4658888	10.67	32.01
1	BX	1	36 CNT	REESES PEANUT BUTTER CUP	1000033	14.15	14.15

Example 4: Using the *System Food Service Distributors* Invoice above, answer the following questions.

a. How many total pounds of sliced peaches did you receive?

There are two *5 LB* cans in each case. You bought 4 cases.
$\Rightarrow$ 2(cans) x 5(pounds each) x 4(cases) = 40 lbs.

b. How much do the peaches cost per pound?

Peaches cost $16.19 per case. Each case has 2 (pack) x 5 (LBS) = 10 lbs. You see that each case, which contains *10 LBS*, costs $16.19. You must divide the total cost by the total pounds to find price per pound.
$\Rightarrow$ $16.19 $\div$ 10 = $1.619 $\cong$ $1.62 per pound.

c. How many 1-ounce servings of granulated sugar can you get from your order?

You ordered two *50 LB* bags of sugar. That makes 100 pounds. You know that there are 16 ounces in one pound
$\Rightarrow$ 100 *pounds* x 16 *ounces* = 1600 ounces of sugar.

d. How much is the butter per pound? Per ounce?

There are 36 pounds of butter in a case as shown on the invoice. The case cost $48.78. That means that 36 pounds of butter costs $48.78. To find the price per pound, divide the total cost by the total pounds.
$\Rightarrow$ $48.78 $\div$ 36lbs = $1.355 $\cong$ $1.36 per pound
$\Rightarrow$ $1.355 $\div$ 16oz = $0.0847 $\cong$ 8¢ per ounce

e. What is the yield in ounces of 3 bags of *GOLDMDL FLOUR*?

The flour comes in a 50-pound bag. Therefore, the yield in ounces is:
50 *pounds* x 16 *ounces* = 800 ounces of flour per bag
3 bags of 800 ounces each = 2400 ounces of flour

f. What is the price per ounce of *GOLDMDL FLOUR*?

The flour comes in a 50-pound bag that costs $10.67.
$\Rightarrow$ $10.67 $\div$ 50 = $0.2134 per pound
$\Rightarrow$ $0.2134 $\div$ 16 $\cong$ $0.013 per ounce

g. If you eat 6 peanut butter cups, then figure the price per cup, how much would each peanut butter cup be worth?

You began with 36 peanut butter cups and ate six. You now have only 30 peanut butter cups. The price of the 36 cups was $14.15. To find the value of the peanut butter cups left, divide the total price by the 30 cups that are left.
$\Rightarrow$ $14.15 $\div$ 30 = $0.4717 $\cong$ 47¢ for each cup.

h. The suggested list price of one peanut butter cup is 55¢. If you sold the entire box of peanut butter cups, how much profit would you make?

If you sold the entire box of 36 peanut butter cups for 55¢ each, you would take in 36 cups x $0.55 each = $19.80. The original box cost you $14.15, so you would make:

REVENUE $-$ COST = PROFIT
$\Rightarrow$ $19.80 $-$ $14.15 = $ 5.65 profit on the whole box.

SAD BUT TRUE

Just think! If you ate six of the peanut butter cups, you would only collect
$\Rightarrow$ 30 x $0.55 = $16.50, which would be a meager $2.35 profit.

Exercise Set 9

For 1–10, mark each problem either True of False.

1. _____All food distribution companies print out price lists each month.

2. _____Manufacturer price lists are the same for all wholesale buyers.

3. _____Many sales reps keep current prices on a laptop computer.

4. _____Food product vendors never quote prices over the phone.

5. _____If a company generates an order guide for your specific business, it will contain all foods that the company has to offer and the prices.

6. _____A requisition is a bill of sale.

7. _____If you own your own small restaurant, you must keep all of your invoices throughout the year. These invoices make up the "cost of goods sold" part of your income statements.

8. _____If you work in a big hotel, you will most likely need to fill out an invoice to get supplies for your kitchen.

9. _____ "# of RU" means the number of recipe units.

10. _____In the pack/size column of an invoice if you see 4/5 LB, it means four-fifths of a pound.

For 11–18, Match each abbreviation.

11. case _____   a.   pc

12. bag_____   b.   ea

13. sack_____   c.   ct or cnt

14. count_____   d.   ctn

15. box_____   e.   sak

16. each_____   f.   bx

17. carton_____   g.   cs

18. piece_____   h.   bg

```
                ORDERING GUIDE LISTING PREVIOUS PURCHASES

Item          Brand     Pack/Size  Description              Cost   Per Yield Price
495 PASTA-FROZEN
    81613     CONTAD    10/1#       FETTUCCINI EGG FRESH FRZN 0.127  OZ  160    ?
625 CATSUP
    80522     HEINZ     24/?OZ      CATSUP BOTTLE              ?     OZ  336  19.71
500 VEGETABLES-CANNED
    55086     NIFDA     6/10#       BEAN CHILI MEXICAN        0.02   OZ  960  14.55
595 JUICES & DRINKS-READY TO USE
    63546     OCEAN     8/60 OZ     CRANBERRY JUICE COCKTAIL   ?     OZ   ?   21.84
    69845     OCEAN     8/60 OZ     ORANGE JUICE 100%          ?     OZ   ?   19.44
    61123     OCEAN     8/60 OZ     GRAPEFRUIT JUICE 100%      ?     OZ   ?   16.81
    67830     NIFDA     12/46 OZ    TOMATO JUICE               ?     OZ   ?   10.36
    65132     NIFDA     12/32 OZ    LEMON JUICE RECONSTITUTED  ?     OZ   ?   15.12
    70005     LIPTON    4/24CT-1OZ  TEA BAGS ICED              ?     PT   ?   19.61
```

For 19–28, round to the nearest tenth of a cent.

19. Find the yield and approximate cost per ounce for the Cranberry Juice Cocktail.

20. Find the yield and approximate cost per ounce for the Grapefruit Juice 100%.

21. Find the yield and approximate cost per ounce for the Orange Juice 100%.

22. Find the yield and approximate cost per ounce for the Tomato Juice.

23. Find the yield and approximate cost per ounce for the Lemon Juice Reconstituted.

24. Find the number of cups and then the cost per cup of iced tea you will be able to make with the Tea Bags Iced. (*One ounce of tea will make 25 cups of iced tea.*)

25. Find the cost per ounce of Catsup.

26. How many ounces of Catsup are in one bottle?

27. Find the price of 10 pounds of Fettuccini Egg Fresh Frzn.

28. Find the price of 15 pounds of Bean Chili Mexican Style.

Challenge Problems

29. If one ounce of raw rice yields 4.88 ounces of cooked rice, find the cost of an 8 oz serving of cooked rice if the price of a 25-pound box of raw rice is $9.80.

30. If 6 ten-inch cherry pies cost $22.57, and the cost of each slice is 47¢, how many slices has each pie been cut into?

For 31 and 32, Use the given blank invoice and complete *all blanks*, including Totals. The ORD column means amount ordered. (*Note: Round the* **COST** *to three decimal places, and all other numbers to two places, or the nearest cent.*)

31. Order No. is 16-44466, Invoice No. is 59130, Invoice date is 6/02/2003, Terms Net 30 Days, Date shipped 6/02/2003, Salesman is Monte M. Your business name is up to you. Give a complete name, address, and phone number.

ITEM #	ORD	PACK/SIZE	DESCRIPTION	PRICE
Item # 2501	3	10#	Croutons Seasoned	$15.60
Item # 3545	2	10#	Bacon Bits-Imitation	$8.75
Item # 3025	1	40/4oz pieces	Beef Liver Sliced	$15.25
Item # 3132	4	2/5#	Chicken Livers	$1.15
Item # 4027	12	24/8oz pieces	Chicken Breast	$42.76
Item # 4522	6	11#/4oz fillets	Walleye Fillet	$85.17
Item # 4524	8	6/3#	Shrimp Frozen	$145.80

Wholesale City Supply

PHONE 808-546-4555

Name

Address

Phone

SALESMAN	DATE SHIPPED	TERMS	INVOICE DATE	INVOICE NO	ORDER NO.

ITEM	ORD	Pack/Size	DESCRIPTION	COST	PER	YIELD	PRICE	EXTENDED AMOUNT
					OZ			
					OZ			
					PC			
					OZ			
					PC			
					PC			
					OZ			

ITEM TOTALS _____

6.98% STATE SALES TAX _____

1% COUNTY SALES TAX _____

32. Order No. is 770-889, Invoice No. is 35476, Invoice date is 3/12/2001, Terms Net 14 Days, Date shipped 3/12/2001, Salesman is Virginia G. Your business name is up to you. Give a complete name, address, and phone number.

ITEM #	ORD	PACK/SIZE	DESCRIPTION	PRICE
Item # 1034	2	4gal tub	Dressing Salad House	$18.60
Item # 2888	1	4/gal	Dressing Buttermilk	$31.08
Item # 2432	6	18/gal	Dressing Mix Bleu Cheese	$45.75
Item # 3414	3	6/#10*	Applesauce	$16.80
Item # 8009	5	115CT** case	Lemons	$18.75
Item # 8675	2	4/5# case	Cabbage Mixed Chopped	$10.43
Item # 8966	8	6/32oz	Horseradish Grnd Prep	$16.38

*#10 is a can size that holds 12 to 13 cups (10# indicates 10 pounds). Use 12 cups for problem 32.

**CT = count

Wholesale City Supply

PHONE 808-546-4555

Name						
Address						
Phone						

SALESMAN	DATE SHIPPED	TERMS		INVOICE DATE	INVOICE NO	ORDER NO.

ITEM	ORD	Pack/Size	DESCRIPTION	COST	PER	YIELD	PRICE	EXTENDED AMOUNT
					OZ			
					OZ			
					OZ			
					OZ			
					EA			
					OZ			
					OZ			

ITEM TOTALS

6.98% STATE SALES TAX

1% COUNTY SALES TAX

PAY THIS AMOUNT

CHAPTER 10

Guest Checks, Tips, and Guestimation

As you know, most fast food restaurants use computer-based ordering techniques, not guest checks. Many of the larger, more exclusive restaurants ask the waitpeople to take orders from memory and then input them into the computerized cash register. Still other establishments use blank guest checks for taking orders. But then you must go to the computerized register, key in the information, and let the computer do all of the mathematics for you. Finally, there are the small, unobtrusive, privately owned cafes and roadhouses that still do everything "by hand." In these service-oriented businesses, the waitperson takes the order from the customer, verbally places the order with the cook, brings the food, clears the dirty dishes, adds the check, adds the tax, collects the money, and has to make change! Wow . . . I'm tired just telling you everything this person has to do for one customer . . . Now that is a person who deserves a good 22% tip! What a refreshing thought!

The computerized method is much better for the business owner in the long run. It alleviates the possibility of calculation errors and makes it much harder for a dishonest worker to steal. Other real advantages the computer system offers are keeping track of all food and drink items sold during the day, the amount of money collected by each server (needed for tax purposes), the specific entrees each server sold, the table numbers each waitperson serviced, as well as the amount of time the customers were seated. The accounting and record-keeping possibilities, especially for tax records, are endless.

Thanks to computers, the necessity of one person's learning to become a skillful guest check author, calculator, tax figurer, or change maker is not quite as important as it once was. Most or all of the arithmetic is done by a computerized cash register. However, the fact remains that not all foodservice operations will be computerized within the next fifteen years.

Last, keep in mind that you need to write legibly, get to know the abbreviations each cook desires, and be cheerful when you take an order.

Artie's Restaurant

LUNCH MENU

Starters		Artie's Favorites	
Artie's Famous Artichoke	Market Price	Fresh Island Fish and Chips	9.95
		Big Island Pork Ribs	9.95
Grilled and Chilled Asian Shrimp	7.95	Baja Fish Tacos	7.95
Crab Wontons	5.50	Roasted Mushroom & Tomato Pizza	6.95
Calamari	6.95	Thai Chicken Pizza	7.95
Chicken Quesadilla	5.95		
Fisherman's Chowder	3.95		

Salads		Desserts	
Crispy Wonton Salad	2.95	Mud Pie	4.95
All You Can Eat Salad Bar	8.95	Double Chocolate Fudge Cake	4.95
Chicken Caesar Salad	8.95	New York Cheesecake	4.95
Island Spiced Fresh Fish Salad	9.95	Tropical Sorbet	3.95
Soup and Salad	6.95		
House Caesar Salad	3.50		

Sandwiches		Beverages	
Huli Huli Sandwich	7.95	Kona Coffee	2.00
Seafood Sandwich	7.95	Decaf Kona Coffee	2.00
Turkey Breast Sandwich	6.95	Hawaiian Spring Water	1.50
Cheeseburger	7.95	Iced Tea	1.50
Veggie Burger	6.95		
Filet of Mahi Mahi Sandwich	8.95		

Example 1:   Look at the menu and write up separate guest check for the following orders.

❖ 4/6/02, 4 persons, separate checks, Table 17, Guest Checks # 1218–1221.

❶ Calamari, Caesar Salad, Turkey Breast Sandwich, Iced Tea.

❷ Chicken Quesadilla, Crispy Wonton Salad, Spring Water.

❸ Salad Bar, Thai Pizza, Decaf, Mud Pie.

❹ Chowder, Fish and Chips, Kona Coffee.

**Plus an order from the bar, Martini, $4.50 for each person, on the back.

a. Make sure that you add the bill up correctly and show the subtotal of the meal.

b. Figure 9.25% food tax and write it on the check.

c. Add the subtotal and tax and enter the sum on the FOOD line. Then add the bar to find the total amount owed. In some states there is a liquor tax paid by the business at the time of purchase that is not taxed again when sold in mixed drinks. The liquor is taxed only when sold as package goods.

Check 1218

Artie's Restaurant
Chicago's Finest Polynesian

Persons	Server	Table #	Date
4	Peter	5	4/6/02

1	Calamari	6.95
1	Caesar Salad	3.50
1	Turkey Sanwich	6.95
1	Iced Tea	1.50

SUBTOTAL	18.90
9¼ % Chicago City Tax	1.75
Food	20.65
Bar	4.50
PAY ONLY THIS AMOUNT	25.15

Check 1219

Artie's Restaurant
Chicago's Finest Polynesian

Persons	Server	Table #	Date
4	Peter	5	4/6/02

1	Quesadilla	5.95
1	Wonton Salad	2.95
1	Sp Water	1.50

SUBTOTAL	10.40
9¼ % Chicago City Tax	0.96
Food	11.36
Bar	4.50
PAY ONLY THIS AMOUNT	15.86

Check 1220

Artie's Restaurant
Chicago's Finest Polynesian

Persons	Server	Table #	Date
4	Peter	5	4/6/02

1	Salad Bar	8.95
1	Thai Pizza	7.95
1	Decaf	2.00
1	Mud Pie	4.95

SUBTOTAL	23.85
9¼ % Chicago City Tax	2.21
Food	26.06
Bar	4.50
PAY ONLY THIS AMOUNT	30.56

Check 1221

Artie's Restaurant
Chicago's Finest Polynesian

Persons	Server	Table #	Date
4	Peter	5	4/6/02

1	Chowder	3.95
1	F & C	9.95
1	Kona Cof	2.00

SUBTOTAL	15.90
9¼ % Chicago City Tax	1.47
Food	17.43
Bar	4.50
PAY ONLY THIS AMOUNT	21.87

d. Figure a 15% tip on the total for the table, so you will know if the party was courteous.

A 15% tip would be 0. 15 × (25.15 + 15.86 + 30.56 + 21.87)
$\Rightarrow 0.15 \times 93.44 = 14.016 \cong \$14.02$

e. Figure a 20% tip on the total for the table, so you will know if the party has class!

A 20 % tip would be 0.20 × (25.15 + 15.86 + 30.56 + 21.87)
$\Rightarrow 0.20 \times 93.44 = 18.688 \cong \$18.69$

f. Figure the change you will need to give each person if each person pays with a crisp $50 bill.

Change for #1218   (50 – 25.15) → $24.85
Change for #1219   (50 – 15.86) → $34.14
Change for #1220   (50 – 30.56) → $19.44
Change for #1221   (50 – 21.87) → $28.13

g. Assume the customers pay the entire amount with one credit card. Figure 2.7% of the total. This is the amount that the credit card company will charge your business to process the purchase. (The amount charged by the credit card company can vary, but will be printed in an agreement with the establishment.)

Credit Card Company 2.7% of $93.44
$\Rightarrow 0.027 \quad x \quad 93.44 \quad = \quad 2.52288 \quad \cong \quad \$2.52$

h. Now, Clear the table. (*Just Kidding.*)

"According to legend, the word "tip" is from an innkeeper's sign 'To Insure Promptness.' If patrons gave a few extra coins, they received their drinks faster." *http://www.eatingpa.com/tipping.html*

TIPS

Here are some guidelines for tipping and tip reporting for tax purposes. Your employer will have a copy of the state rules that pertain to your employment. Keep in mind that tips are a major part of a waitperson's compensation.

People in the service industry can be among the hardest working people you will ever meet. Most of them are paid an hourly wage, but they depend on tips to make a living. These are standard guidelines for tipping.

- In restaurants, a tip is based on the service, not the food. You are not tipping the chef, you are tipping the waiter/waitress.

- For good service, you should leave 15% of the total bill. If you are using a coupon or other discount, the tip should be based on the full bill, not the discounted amount.

- If the service was outstanding, 20% is a great way to show it. Alternately, if the service was below par, 10% or lower sends the same message. (Canada website on tipping)

"The underreporting of gratuities received by service professionals in the restaurant industry has been a major focal point by the Internal Revenue Service in recent years. In most instances, it is at the discretion of the service professional to declare the dollar amount of gratuities received that will eventually aggregate to his/her W-2. In the recent age of credit card and debit card payments, any gratuity charged is automatically included as part of an employee's wages and earnings. However, in the case of restaurants where cash payment is ordinary, for example, a college campus, the employee is solely responsible for declaring gratuities in good faith.

Reversing an Alabama district court, the 11th Circuit has held that the IRS has the authority under IRC Sec. 3121(q) to assess a restaurant with the employer's share of FICA taxes computed on an aggregate basis without determining the amount underreported by individual tipped employees and crediting their wage history accounts.

While the court agreed with the restaurant's contention that it is the employees responsibility to ensure they are properly credited for all wages by accurately reporting all tips received, it went on to say that "basing the employer's share of FICA taxes exclusively on employees' reported tips would provide incentive to the employer to discourage reporting or ignore blatantly inaccurate reporting by the employees so that the employer could pay less FICA tax."

Copyright ©1999 Winningham Becker & Company
Last modified: February 19, 1999 http://www.wbac.com/Articles/restfica1.htm

337.065 Unlawful for employer to require remittance of gratuity—Tip pooling.

(1) No employer shall require an employee to remit to the employer any gratuity, or any portion thereof, except for the purpose of withholding amounts required by federal or state law. The amount withheld from such gratuity shall not exceed the amount required by federal or state law.

(2) As used in this section, "gratuity" means voluntary monetary contribution received by an employee from a guest, patron, or customer for services rendered.

(3) No employer shall require an employee to participate in a tip pool whereby the employee is required to remit to the pool any gratuity, or any portion thereof, for distribution among employees of the employer.

(4) Employees may voluntarily enter into an agreement to divide gratuities among themselves. The employer may inform the employees of the existence of a voluntary pool and the customary tipping arrangements of the employees at the establishment. Upon petition by the participants in the voluntary pool, and at his own option and expense, an employer may provide custodial services for the safekeeping of funds placed in the pool, if the account is properly identified and segregated from his other business records and open to examination by pool participants.

*Effective: July 15, 1996 **History:** Amended 1996 Ky. Acts ch. 115, sec. 2, effective July 15, 1996.*
Created 1976 Ky. Acts ch. 222, sec. 1. http://www.lrc.state.ky.us/krs/337-00/065.PDF

TIPPED EMPLOYEES—Effective July 15, 1998, for any employee engaged in an occupation in which more than $30 dollars per month is customarily and regularly received in tips, the employer may pay a minimum of $2.13 per hour if the employer's records can establish for each week where credit is taken, when adding the tips received to wages paid, not less than the minimum wage is received by the employee. Subsequently, the tipped rate will adjust in accordance with the federal minimum tipped rate as prescribed by 29 U.S.C. Sec. 206(a)(1). No employer shall use all or part of any tips or gratuities received by employees toward the payment of the minimum wage. (KRS 337.275(2)) No employer shall require an employee to remit to the employer any gratuity, or any portion thereof, except for the purpose of withholding amounts required by federal or state law. No employer shall require an employee to participate in a tip pool whereby the employee is required to remit to the pool any gratuity, or any portion thereof, for distribution among employees of the employer. Employees may voluntarily enter into an agreement to divide gratuities among themselves. The employer may inform the employees of the existence of a voluntary pool and the customary tipping arrangements of the employees at the establishment. Upon petition by the participants in the voluntary pool, and at the employer's own option and expense, an employer may provide custodial services for the safekeeping of funds placed in the pool if the account is properly identified and segregated from the other business records and open to examination by pool participants. (KRS 337.065)

RECORDS—Every employer subject to the provisions of the Kentucky Minimum Wage Law shall make and preserve records containing the following information: (a) Name and address of each employee; (b) Hours worked each day and each week by each employee; (c) Regular hourly rate of pay; (d) Overtime hourly rate of pay for hours in excess of forty hours in a workweek; (e) Additions to cash wages at cost, or deductions (meals, board, lodging, etc.) from stipulated wages in the amount deducted, or at cost of the item for which deductions are made; (f) Total wages paid for each workweek and date of payment. Such records shall be kept on file for at least one year after entry. No particular form or order is prescribed for these records provided that the information required is easily obtainable for inspection purposes. (KRS 337.320)

KENTUCKY WAGE AND HOUR LAWS POST THIS ORDER WHERE ALL EMPLOYEES MAY READ KRS 337.200) http://www.state.ky.us/agencies/labor/labrhome.htm

City and County of Denver TAX GUIDE
Department of Revenue
Treasury Division
Topic No. 77

TIPS AND GRATUITIES

Tips and/or gratuities are not subject to sales and use tax when the patron voluntarily leaves cash or adds an amount to the charge ticket for the benefit of the service provider. However, mandatory service or service related charges, whether described as tips, gratuities, or otherwise, that are included as part of the amount paid for food or drink, served by restaurants and at other places at which prepared food or drink is regularly sold, are taxable at the rate of 4%. In order for a proprietor-determined service charge to *not* be subject to Denver sales tax, both of the following conditions must apply. First, the charge must be labeled on the customers' guest check as either "suggested" or discretionary" and as a "tip" or a "gratuity." Second, the "total" line on the credit card receipt, where applicable, must be left blank, to enable the customer to decide what amount will actually be paid. Under these conditions, the gratuity is considered to be voluntary, and thus is not subject to Denver sales tax. The difference between Denver's and the State's treatment of gratuities is that if the gratuity is mandatory, it is subject to Denver tax, but is not subject to State tax if distributed to the service person.

EXAMPLES

1. Edgar and Bea go out to a neighborhood restaurant for dinner. When they receive the bill Edgar voluntarily leaves a tip for the outstanding service they received. The tip is not subject to the imposition of sales tax.

2. Nine co-workers go to lunch at a restaurant. The restaurant imposes a mandatory 15% gratuity for parties of six or more. Such a statement appears on the menu and the customers' bill. This gratuity is distributed directly to the waiter or waitress who actually rendered the service. The mandatory 15% gratuity is subject to Denver tax.

3. Twenty co-workers go to a retirement luncheon at a favorite restaurant. The restaurant, as is its practice for parties of six or more, prints a line on the guest check labeled "15% Suggested Gratuity" and includes the 15% in the total amount due on the quest check. However, on the credit card receipt, the total line is left blank to enable the patron to decide what amount will actually be paid. This suggested gratuity is considered voluntary and is thus not subject to the imposition of Denver tax.

THE ABOVE INFORMATION IS A SUMMARY IN LAYMAN'S TERMS OF THE RELEVANT DENVER TAX LAW FOR THIS INDUSTRY OR BUSINESS SEGMENT. IT IS NOT INTENDED FOR LEGAL PURPOSES TO BE SUBSTITUTED FOR THE FULL TEXT OF THE DRMC AND APPLICABLE RULES AND REGULATIONS.

Confused? Rightfully so. There are many laws governing the reporting of tips, so be sure you know your city and state laws.

GUESTIMATION (A LITTLE GUESS . . . A LITTLE ESTIMATION)

Now, back to mathematics. Here is a simple way to figure the amount of tip one should leave or receive, as the case may be. To figure 10% of any number, simply move the decimal point one place to the left. (*You are actually dividing by 10 or multiplying by 1/10.*)

Study the examples below. Notice that I just *dropped the last digit, I did not round off.* This method is for guestimation only—when you just want to be close, and do the figuring in your head.

10% of $13.42 $\Rightarrow$ $1.34**2** $\cong$ $1.34   10% of $167.90 $\Rightarrow$ $ 16.79

10% of $ 5.75 $\Rightarrow$ $0.57**5** $\cong$ 57¢   10% of $82.38 $\Rightarrow$ $8.23**8** $\cong$ $8.23

Ah, that was easy. Now try to find a guestimation of 20% of the given numbers. Do you have any ideas?

You probably know that 20% is just *twice* 10%. So, after you have guestimated 10% of any number, you simply multiply by 2 to guestimate 20%. *Remember, this is not exact!*

Think

20% of $13.42   10% of $13.42 $\cong$ $1.34 $\Rightarrow$ $\times$ **2** $\cong$ $2.68

20% of $5.75   10% of $ 5.75 $\cong$ 57¢ $\Rightarrow$ $\times$ **2** $\cong$ $1.14

As long as you are guestimating, when you do a harder one, you may want to round up or down to the *nearest whole number* before you multiply by two.

20% of $167.90 $\Rightarrow$ 10% of $167.90 $\cong$ $16.79 × 2

$\qquad\qquad\qquad\qquad\qquad\qquad\qquad$ *think* $\qquad$ **$17** x 2 $\Rightarrow$ $34

20% of $82.89 $\Rightarrow$ 10% of $82.89 $\cong$ $8.28 x 2

$\qquad\qquad\qquad\qquad\qquad\qquad\qquad$ *think* $\qquad$ **$8** × 2 $\Rightarrow$ $16

Finally, let's try the standard tip of 15%. You know that 5% is *half* of 10%, so to find 15% of any number, just add 10% of the number + 5% of the number. Sounds hard, but as long as you are guestimating, it isn't that bad.

It is difficult to take half of a four-digit number in your head, or to take half of any odd number in your head. So this time, round up or down to an even *whole number dollar amount* before taking 10%. Watch:

Example 2: Using guestimation, show how you might find 15% of $86.96 in your head. *Remember, it is only an approximation, so round up to $90.00.*

Step 1: Find 10% of $90.00 in your head.

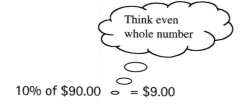

10% of $90.00 = $9.00

Step 2: Now find 5% of the number using your guestimation for 10%.

5% of $90 is just half of 10% $\Rightarrow$ 1/2 of $9.00 = $4.50

Step 3: Finally, add 10% of the number + 5% of the number without paper and pencil.

9.00 + 4.50 = $13.50

Rounding to an even dollar amount is by far the easiest for division purposes. That is why I rounded $86.95 up to $90. Had I rounded up to $87, I would have had to take half of 87 in my head! *No, thanks.* Why make it hard on yourself if you are only trying to guestimate?

Here are a few more to study. Enjoy!

15% of $24.75 $\Rightarrow$ 10% of 24 = 2.40 *Round down to 24 for easy division.*

$\Rightarrow$ 5% of 24 = 1/2 of 2.40 = 1.20

$\Rightarrow$ 2.40 + 1.20 = $3.60

15% of $13.45 $\Rightarrow$ 10% of 14 = 1.40 *Round up to 14 for easy division.*

$\Rightarrow$ 1/2 of 1.40 is 0.70 $\Rightarrow$ 1.40 + 0.70 $\Rightarrow$ $2.10

15% of $157.51 $\Rightarrow$ 10% of 160 = 16.00 *Round up to 160.*

$\Rightarrow$ 1/2 of 16 is 8 $\Rightarrow$ 16 + 8 $\Rightarrow$ $24.00

O.K., You want to know how close your guestimations were? Let's check.

15% of 86.96	= 13.044	$\cong$ $13.04	guessed $12.90
15% of 24.75	= 3.7125	$\cong$ $ 3.71	guessed $ 3.60
15% of 13.45	= 2.0175	$\cong$ $ 2.02	guessed $ 2.10
15% of 157.51	= 23.625	$\cong$ $23.63	guessed $24.00

As you can see, the guestimation process that you did *in your head* is acceptable for figuring a tip. It also comes in handy when you are shopping and see a sticker that reads "15% off. " You will know the price *before* you check out and be able to monitor the bill.

More and more I witness scanning devices that "make mistakes" on sale items. Whether or not the mistakes are intentional, you will be aware of what your purchase should cost and feel confident.

Now it is your turn to practice.

1. What is 15% of 400?   10% is ?   Take half of that. Add.   So 15% of 400 is?

2. What is 15% of 66?   10% is ?   Take half of that. Add.   So 15% of 66 is?

3. What is 20% of 80?   10% is ?   Take twice that amount.   So 20% of 80 is?

4. What is 20% of 42?   10% is ?   Take twice that amount.   So 20% of 57 is?

Answers: 1. 40, 20, 60; 2. 6.6, 3.3, 9.9; 3. 8, 16; 4. 4.2, 8.4.

Exercise Set 10

For dinners at Artie's Restaurant assume a 7% sales tax.

    a.   Fill out completely one separate guest check for questions 1–6 below.

    b.   Figure a 15% tip on the total for the table, so you will know if the party was courteous.

    c.   Figure a 20% tip on the total for the table, so you will know if the party has class!

    d.   Figure the change you will need to give if the table pays with a crisp hundred dollar bill.

    e.   Assume the customer pays with a credit card. Figure 2.3% of the total. This is the amount that the credit card company will charge your business to process the purchase.

1.   a.   ❖ 1/9/02, 3 persons, Table 1,
           Guest Check # 90210
          ❶ Pork Ribs, Wonton Salad,
            Spring Water
          ❷ Fish and Chips, Decaf Coffee
          ❸ Chowder, Island Fish Salad,
            Sorbet, Coffee

    b.   15% tip _____

    c.   20% tip _____

    d.   Change from $100? _____

    e.   Credit charge _____

Artie's Restaurant

No._____

Persons	Server	Table #	Date

SUBTOTAL _____

Tax _____

PAY ONLY THIS AMOUNT _____

2. a. ❖ 4/8/99, 2 people, Table 16,
      Guest Check # 4824
      ❶ Calamari, Mushroom Pizza, Fudge Cake,
      Decaf
      ❷ Thai Pizza, Salad Bar, Iced Tea

   b. 15% tip _____

   c. 20% tip _____

   d. Change from $100? _____

   e. Credit charge _____

	No._____
Artie's Restaurant	

Persons	Server	Table #	Date

SUBTOTAL	_____
Tax	_____
PAY ONLY THIS AMOUNT	_____

3. a. ❖ 7/4/01, 4 persons, Table 8,
      Guest Check # 1001
      ❶ Huli Huli Sandwich, House Caesar Salad,
      Sorbet, Iced Tea
      ❷ Asian Shrimp, Veggie Burger, Mud Pie, Iced
      Tea
      ❸ Asian Shrimp, Fish Tacos, Wonton Salad,
      Sorbet, Iced Tea
      ❹ Mahi Sandwich, Mud Pie, Iced Tea

   b. 15% tip _____

   c. 20% tip _____

   d. Change from $100? _____

   e. Credit charge _____

	No._____
Artie's Restaurant	

Persons	Server	Table #	Date

SUBTOTAL	_____
Tax	_____
PAY ONLY THIS AMOUNT	_____

4. a. ❖ 11/22/00, 3 persons, Table 12,
   Guest Check # 1122
   ❶ Crab Wontons, Soup and Salad, Sorbet, Decaf
   ❷ Quesadilla, Cheeseburger, Cheesecake
   ❸ Calamari, Fish and Chips, Iced Tea

   b. 15% tip _____

   c. 20% tip _____

   d. Change from $100? _____

   e. Credit charge _____

		No._____	
	Artie's Restaurant		
Persons	Server	Table #	Date
	SUBTOTAL		_____
	Tax		_____
	PAY ONLY THIS AMOUNT		_____

5. a. ❖ 2/7/03, 3 persons, Table 6, Guest
   ❶ Check # 27203
   ❷ Calamari, House Caesar Salad, Thai Pizza, Decaf Coffee
   ❸ Cheeseburger, Decaf Coffee
   ❹ Shrimp, Pork Ribs, Iced Tea

   b. 15% tip _____

   c. 20% tip _____

   d. Change from $100? _____

   e. Credit charge _____

		No._____	
	Artie's Restaurant		
Persons	Server	Table #	Date
	SUBTOTAL		_____
	Tax		_____
	PAY ONLY THIS AMOUNT		_____

6. a. ❖ 9/16/04, 3 persons, Table 16,
        Guest Check # 91604
        ❶ Crab Wontons, Fish and Chips, Sorbet,
        Decaf
        ❷ Chowder, Mushroom Pizza, Sorbet,
        Iced Tea
        ❸ Crab Wontons, Fish and Chips,
        Iced Tea

    b. 15% tip _____

    c. 20% tip _____

    d. Change from $100? _____

    e. Credit charge _____

	No. _____
	Artie's Restaurant

Persons	Server	Table #	Date

SUBTOTAL	_____
Tax	_____
PAY ONLY THIS AMOUNT	_____

7. Using guestimation, figure 10% of $14.65.

8. 10% of any number can be found simply by (explain) _____?_____

    a. Once you have guessed what 10% of a number is, how would you find 5% of that same number?

    b. How would you guestimate 15% of any number?

9. If 10% of a given number is $6.22, what is 5% of the given number?

    15% of the given number?

    20% of the given number?

10. Guestimate. Find 10%, then round to the nearest whole number before multiplying by two. Show your thinking.

    a. 20% of $ 1,243.56 *think* ⇒ *124.3* ⇒ *124* ⇒ *x2* ⇒ *$248*

b. 20% of 163.48

c. 20% of 400

d. 20% of $ 46.99

e. 20% of $ 8.73

f. 20% of 37

11. Find the answers to question 10 using a calculator. Round to the nearest cent.

a. 20% of $ 1,243.56

b. 20% of 163.48

c. 20% of 400

d. 20% of $ 46.99

e. 20% of $ 8.73

f. 20% of 37

12. Guestimate. If necessary, round to an *even whole number*, before taking 10%. Try to do all of the figuring in your head, including the addition. Show your thinking.

a. 15% of 25.70 *think* $\Rightarrow$ 26 $\Rightarrow$ 2.6 $\Rightarrow$ *half of that is* 1.3 $\Rightarrow$ *add* $\Rightarrow$ 3.9

b. 15% of 5000

c. 15% of $ 841.44

d. 15% of 6.50

e. 15% of 600

f. 15% of $ 47.32

13. Find the answers to question 12 with a calculator. Round to the nearest cent.

    a. 15% of 25.70

    b. 15% of 5000

    c. 15% of $ 841.44

    d. 15% of 6.50

    e. 15% of 600

    f. 15% of $ 47.32

14. How would you find 30% of $43.66 without a calculator? Explain your steps.

15. See how many of the following problems you can do in your head, writing down only the answer, in 1 minute! Ready . . . Set . . . GO!

    a. 10% of 6000   _____

    b. 10% of 50   _____

    c. 5% of 800   _____

    d. 10% of 537   _____

    e. 15% of 20   _____

    f. 15% of 400   _____

g. 20% of 75   _____

h. 30% of 90   _____

i. 1 % of 30   _____

j. 40% of 20   _____

k. 10% of 45.8   _____

l. 15% of 6000   _____

m. 10% of 50,000   _____

n. 20% of 700   _____

CHAPTER 11

Paychecks and Income Statements

PAYCHECKS

If you have already held down a job, you are probably familiar with the idea that you don't usually take home all of the money that you earn. The money that you earn is called your *gross income*.

The actual money that you take home is called your *net pay*. If you work by the hour, you earn a *wage*. If the amount of money you earn is agreed upon for the year, or sometimes for the month, this is called a *salary*.

Often management positions are salaried positions. You are expected to work as long as it takes to get the job done right even if it means many more than 40 hours per week. In other words, in a salaried position, you are paid the amount agreed upon per month, no matter how many hours you work. This is quite different than a wage earner who gets paid by the hour, with overtime pay for working more than 40 hours per week.

It is important to understand every part of your paycheck, so that you can be sure the bookkeeper has not made any mistakes. Remember, although payroll is done on a computer in many instances, a real person had to punch in the data used to figure your pay. Be aware at all times of how much your paycheck should be and inspect it carefully.

Words of Wisdom

*No one will have **your** best interest in mind except **YOU**. You must realize that only you will expend every ounce of energy you have to take care of your family. You and you alone must be responsible for what happens to you.*

With that in mind, learn how to examine your paycheck and understand every little detail. Inspect and reconcile your bank statements. Be sure that every aspect of your financial well-being is considered and analyzed. If you know and understand percents, payroll deductions, interest, sales prices, taxes, mortgages, you will be confident, not be intimidated by figures. You will know before you sign on the dotted line just how much interest you will actually have to pay for a loan. You alone will be in control of your life, your finances, and your future.

```
Arties Restaurant Corporation                                              23801

EMPLOYEE NANE:    VALERIE STATHAM          EMPLOYEE ID:    Hula Girl
SOCIAL SEC. #:    505-67-1234 RATE REG:  8.50 OT: 0.00  CHECK NUMBER:  23888
PAY PERIOD:       11/16/01 - 11/30/01                   CHECK DATE:     12/07/01

REG: HOURS:  54.50 PAY        463.25   Tips  Deducted - Lunch          603.00 -
O/T: HOURS     0.00 PAY         0.00   OTHER DEDUCTION 2                 0.00
Tips Reported-Lunch            603.00  OTHER DEDUCTION 3                 0.00
                                       OTHER DEDUCTION 4                 0.00

----- GROSS PAY -----         1066.25
FEDERAL INCOME TAX:            106.62 -
FICA:                          78.60 -  ----- TOTAL WITHOLDING ----     863.03 -
STATE INCOME TAX:              74.81 -
LOCAL INCOME TAX:               0.00   ------ NET PAY ------            203.22

-----------------------YEAR - TO - DATE -----------------------------
      GROSS PAY FED TAX   FICA TAX   STATE TAX LOCAL TAX   OTHER W/H   NET PAY
      23,096.93  2,449.55  1,702.62  1,675.81    0.00      13,110.49   4,158.46
```

The paycheck stub above reflects a business in Hawaii. Notice there is no local tax. Also notice that the tips were reported and then deducted. Although the tips are figured as income in the GROSS PAY line to be paid by the employer, the tip amount is then deducted prior to figuring the NET PAY. This way the tips are reflected in your gross pay, from which all federal and state taxes are figured.

Your net pay, also called take-home pay, is the dollar amount that you actually get to take home. Your net pay is your gross pay minus all deductions. Deductions may include health benefit premiums; union dues; state, federal, and local federal taxes; FICA, child support payments; and many others.

When reading a pay stub from an employer that has had to deduct tips, the gross pay may look deceivingly low. However, keep in mind that Valerie has already "taken home" the $603 in tips. Thus, although her net pay reflects only $203.22, she has actually taken home $806.22.

Valerie's pay rate is $8.50 per hour. The check number, pay period, check date and employee ID are self-explanatory. REG hours, or regular hours, refers to hours, worked by Valerie at the "regular rate," which in this case is $8.50 per hour.

$$54.50 \text{ hours @ } 8.50 \text{ per hour} \Rightarrow 54.50 \bullet 8.50 = \$463.25$$

O/T or overtime hours are listed as 0. The gross pay, then, is the sum of the regular hourly pay wage plus the reported tips.

$$\$463.25 + \$603.00 = \$1,066.25 \text{ GROSS PAY}$$

Valerie's net pay is then figured by deducting tips, federal and state income taxes, plus of course FICA, the Federal Insurance Contributions Act. There is a federal law that requires your employer to withhold two separate taxes from your wages: a social security tax and a Medicare tax. Together these make up FICA contributions.

Let us take a closer look at Valerie's tax deductions.

Example 1:   What percent of Valerie's gross pay is the federal income tax?

Recall from Chapter 5, the concept of the part *to the* whole.

$$\frac{part}{whole} = \frac{106.62}{1,066.25} \cong 0.0999 \cong 10\%$$

Valerie pays 10 % of her gross pay in federal income taxes.

Example 2:   What percent of Valerie's gross pay is her state income tax?

$$\frac{part}{whole} = \frac{74.81}{1,066.25} \cong 0.0701 \cong 7.01\%$$

Valerie pays 7.01% of her gross pay in state income taxes.

Example 3:   What percent of Valerie's "year to date" gross pay is her FICA tax?

$$\frac{part}{whole} = \frac{1,702.62}{23,096.93} \cong 0.0737 \cong 7.37\%$$

The tips made are in the *OTHER DEDUCTION 1* spot on Valerie's paycheck stub. At the bottom in the *YEAR-TO-DATE* section, the year's accumulated tip deductions are listed under *OTHER W/H*, or other withholding.

Example 4:   If Valerie made $1,200 in tips this month, what percent of her year-to-date tip total would this month's tips be?

$$\frac{part}{whole} = \frac{1,200}{13,110.49} \cong 9.15\%$$

This month's tips would be 9% of her tips so far this year.

Example 5:   What percent of Valerie's gross pay this month is her FICA tax?

$$\frac{part}{whole} = \frac{78.60}{1066.25} \cong 0.0737 \cong 7.37\%$$

INCOME STATEMENTS

On the next page you will find an Income Statement for the year 2000. This is an actual full year's accounting for a real restaurant. The name has been omitted, but the figures remain true. There is a wealth of information on this one page. Let's see if you can figure a bit of it out.

Income Statement

For the Twelve Months Ending December 31, 2000

Revenues	Current Month	%	Year to Date	%
Restaurant	45,214.03	65.13	849,901.78	74.4
Bar	16,288.00	23.46	256,616.45	22.46
Video Games	7,915.00	11.4	35,860.00	3.14
Total Revenues	69,417.03	100	1,142,378.23	100
Cost of Sales				
Cost of Goods Sold	(22,312.91)	(32.14)	373,609.77	32.70
Total Cost of Sales	(22,312.91)	(32.14)	373,609.77	32.70
Gross Profit	47,104.12	67.86	768,768.46	67.30
Expenses				
Gross Wages	68,287.18	98.37	359,753.87	31.49
FICA Tax Expense	5,939.74	8.56	44,096.40	3.86
FUTA Tax Expense	1,242.20	1.79	1,633.89	0.14
SUTA Tax Expense	12,833.77	18.49	21,149.77	1.85
Employee Benefits	(2,751.71)	(3.96)	31,718.15	2.78
Rent	0.00	0.00	42,648.13	3.73
Utilities	1,344.82	1.94	19,681.81	1.72
Telephone	0.00	0.00	851.10	0.07
Video Game Expenses	1,303.40	1.88	5,514.79	0.48
Advertising & Promotion	55.83	0.08	6,596.63	0.58
Repairs & Maintenance	1,605.22	2.31	8,702.11	0.76
Operating Supplies	1,153.40	1.66	6,274.86	0.55
Office Supplies	0.00	0.00	482.87	0.04
Postage	0.00	0.00	84.00	0.01
Interest Expense	(40,277.24)	(58.02)	10,901.21	0.95
Bank Charges	1.39	0.00	150.90	0.01
Credit Card Discounts	411.37	0.59	7,639.30	0.67
Legal & Accounting	704.75	1.02	6,138.25	0.54
Insurance	5,378.59	7.75	21,976.34	1.92
Auto Expenses	497.00	0.72	11,829.88	1.04
Dues & Subscriptions	0.00	0.00	2,335.00	0.20
Permits & Licenses	25.00	0.04	843.00	0.07
Contributions	0.00	0.00	575.00	0.05
Sales Taxes	2,961.22	4.27	58,597.91	5.13
Personal Property Taxes	0.00	0.00	250.00	0.02
Real Estate Taxes	0.00	0.00	14,425.27	1.26
Depreciation - Furn & Equip	13,456.31	19.38	13,456.31	1.18
Depreciation - Leasehold	1,391.42	2.00	1,391.42	0.12
Miscellaneous Expenses	0.00	0.00	226.85	0.02
Total Expenses	75,563.66	108.85	699,925.02	61.24
Net Income	(28,459.54)	(40.99)	68,843.44	6.06

Revenues

1. What was the total revenue for the business for the year 2000?

2. What percent of the total revenue did the Video Games make up?

3. What percent of the total revenue was made in the Bar?

4. What percent of the total revenue did the Restaurant bring in?

The answers to these questions are already figured for you. Just look in the last column of the statement, under the % sign for 2–4, and in the third column under Year to Date for question 1.

Cost of Sales

This section lists the cost of the raw food sold. Notice the numbers are in parentheses to indicate an expense, as opposed to an income. The first listing "Cost of Goods Sold" is (22,312.91) or an expenditure of $22,312.91. This is the amount of money that was spent purchasing the food that was served in the restaurant for the current month of December.

Next to the Cost of Goods Sold listing is a % , also in parentheses, indicting expenditure. This number indicates that 32.14% of the Total Revenue for the month was spent on the raw food costs.

$$\frac{raw \; food \; cost}{total \; revenue} = \frac{22{,}312.91}{69{,}417.03} \cong 0.321432$$

The Gross Profit is figured by: **Gross Profit = Total Revenue − Cost of Sales**

As you can see, the *Gross Profit* for the month of December was $91,729.94. This number was found by subtracting the cost of the raw food from the total revenue.

Total Revenue − Cost = Profit

$69,417.03 − $22,312.91 = $47,104.12

To find the *Net Income* for the month of December, look at the bottom of the page and notice that the expenses totaled more than the Gross Profit. Thus there was a Net Loss for the month of $28,459.54. Further, the *Total Expenses* were 108.87% of the Total Revenue for the month!

$$\frac{expenses}{total \; revenue} = \frac{75{,}563.66}{69{,}417.03} \cong 1.0885 \text{ or } 108.85\,\%$$

Although for the month of December they showed a loss, the *Net Profit* for the year was $68,843.44. This was only 6.06% of the Total Revenues for the year 2000. Not much of a profit margin.

Net Income = Gross Profit − Total Expenses

$68,843.44 = $768,768.46 − $699,925.02

Finally, let's look at some of the expenses for the business.

1. Utilities—$19,681.81 for the year. Although that seems like an outrageous amount, you should notice that it is only 1.72 % of the total revenues for the year.

2. Gross Wages—$359,753.87 for the year.

$$\frac{gross\ wages}{total\ revenue} = \frac{359,753.87}{1,142,378.22} \cong 0.3149 = 31.49\%$$

The wages paid out was the largest expense, a close second to the raw food costs. Compare. The cost of the food purchased to cook and sell was 32.7% of the total revenues. The cost of the help to prepare, serve, and clean up was 31.49% of the total revenues.

3. Video Game Expense—$5,514.79 for the year. Consider that the video game room brought in $35,860 as listed in the top part of the income statement. In the expenses we note that the cost to maintain the video games was $5,514.79. *Hmmm* . . . Now since **Profit = Revenue − Cost**, we can say that the video game room made a profit of:

$$\$35,860.00 - \$5,514.79 \Rightarrow Profit = \$30,345.21$$

Exercise Set 11

Use the pay check stub for Valerie Statham, given on page 120, to complete the questions 1–10. Round all percent answers to the nearest hundredth of a percent.

1. What percent of Valerie's year-to-date gross pay is the year-to-date federal tax?

2. If the FICA tax rate is 7.37%, how much FICA tax would be deducted from Valerie's pay if her gross pay for a given month was $985.00?

3. What percent of Valerie's year-to-date gross pay is the year-to-date state income tax?

4. What percent of Valerie's gross pay is the local income tax?

5. You noticed that $92.35 was deducted from your gross pay for state income tax. If your gross pay was $1,358, what percent of your gross pay was withheld?

6. What percent of Valerie's hourly wages for the month ($463.25) is her net pay for the month?

7. What percent of Valerie's year-to-date gross pay is the year-to-date Other Withholding (total tips for the year to date) ?

8. What percent of Valerie's gross pay for the month are the tips?

9. What percent of Valerie's year-to-date tips, *OTHER W/H*, are her tips this pay period?

10. What percent of her year-to-date gross pay is her current pay period gross pay?

Challenge Problem

11. Valerie's regular pay rate has changed to $9.00 per hour. Fill in her paycheck stub. Her *Tips Reported* and *Tips Deducted* were $850. Assume a 7.01% state tax rate. She worked 68.5 hours this time. Federal tax deduction 10%, FICA rate 7.37%, no local tax.

```
Artie's Restaurant Corporation
                                                                    23802
EMPLOYEE NAME:    VALERIE  STATHAM              EMPLOYEE ID:   Hula Girl
SOCIAL SEC. #:      505-67-1234  RATE REG:      9.00    CHECK NUMBER:  23888
PAY PERIOD:         12/16/01  - 12/30/01                CHECK DATE:    1/07/02

REG: HOURS:   68.5 PAY                   Tips  Deducted  - Lunch                 -
O / T: HOURS      0.00 PAY   _____   OTHER DEDUCTION 2                 0.00
Tips Reported -Lunch         _____   OTHER DEDUCTION 3                 0.00
                                         OTHER DEDUCTION 4                 0.00
---- GROSS PAY -----
FEDERAL INCOME  TAX:         _____-
FICA:                        _____-  -------- TOTAL WITHOLDING ----   _____-
STATE INCOME  TAX:           _____-
LOCAL INCOME  TAX:           _____-  ----------------- NET PAY ------  _____
```

Use the Income Statement given in this section for questions 12–26.

12. During the month of December, the business lost money. The expenses were more than the revenues. How much more?

13. For the month of December, what percent of the Video Game revenues were the Video Games expenses?

14.   What percent of the total Video Game expenses for the year occurred during the month of December?

15.   What percent of the total year's revenues made by the Bar were made in December?

16.   How much more was made by the Restaurant than was made by the Bar in December?

17.   The Total Expenses for December were what percent of the Total Expenses for the whole year?

18.   The Total Expenses for December were what percent of the Total Revenues for the whole year?

19.   The Cost of Goods Sold in December were what percent of the year's Total Cost of Goods Sold? One month is what percent of one year?

20.   The Furniture and Equipment Depreciation is considered an expense. On how many Current Month's Ending Statements is this expense listed each year? Explain your answer.

21.   Were the Personal Property Taxes paid in December? Explain.

22.   Were there any contributions made in December?

23. What percent of the Total Advertising Expense for the whole year was spent on Advertising in December?

24. If you took the yearly Rent and divided it into 12 equal payments, how much rent would be due each month?

25. Pretend that this is your business. Of the expenses listed, which ones would you try to increase? Which ones do you feel are not necessary?

26. List in order from greatest to least, the six largest Expenses for the business in the year 1999.

UNIT 4

Food Service Math

CHAPTER 12

Converting Weights and Measures

You have often heard the word "conversion," but do you remember the first time you actually had to convert any quantity from one unit into another? Was it that long ago?

Let's think back to grade school. You were taught that 10 pennies equaled 1 dime; 20 nickels a dollar; 12 inches 1 foot; 2 feet 1 yard; and so on. In the food service industry, there will be many times when you will need to convert quantities from one unit to another.

Here is a chart that should help you with many of these conversions. Although there are computers and calculators that will convert units for you, most of us do not have these calculators readily available.

Equivalent Weights and Measures for the U.S. and Metric Systems

1 cup = 0.24 liters 1 cup = 16 tablespoons 1 cup = 240 milliliters 1 cup = 8 ounces	1 pint = 0.477 liters 1 pint = 2 cups 1 pint = 477 milliliters
1 gallon = 4 quarts 1 gallon = 3.79 liters	1 pound = 0.453 kilograms 1 pound = 16 ounces 1 pound = 454 grams
1 gram = 0.0352 ounces 1 gram = 1,000 milligrams	1 quart = 0.946 liters 1 quart = 2 pints 1 quart = 32 ounces 1 quart = 4 cups 1 quart = 946.3 milliliters
1 kilogram = 1,000 grams 1 kilogram = 2.2 pounds	
1 liter = 0.2642 gallons 1 liter = 1,000 milliliters 1 liter = 1.057 quarts	1 tablespoon = 15 milliliters 1 tablespoon = 3 teaspoons 1 teaspoon = 5 milliliters
1 ounce = 1/8 cup 1 ounce = 2 tablespoons 1 fl. ounce = 29.58 milliliters 1 ounce = 28.35 grams	4 pecks = 1 bushel 8 quarts = 1 peck 1 milliliter = 1 cubic centimeter

Volume: milliliter, teaspoon, tablespoon, liquid ounce, cup, pint, liter, quart, gallon.

Weight: milligram, gram, ounce, kilogram, pound.

Have you ever wondered, "How much does a cup of water weigh?" In the U.S. Customary Measurement System there is not an easy way of remembering the relationships between volume or capacity and weight. For instance, how much does a cubic foot hold? You know, a container with 12-inch sides filled with water. And how much does the water inside it weigh? What are the dimensions of a cubic container that holds a gallon? And how much does a gallon of water weigh?

This is a picture of a **5 cc** syringe next to **5 c**ubic **c**entimeters. One *cubic centimeter* holds *1 milliliter* of water. The ml of water weighs *1 gram*.

The container (top left) is *1 cubic decimeter, (10x10x10=1,000cm³)* which holds *1 liter of water.* A liter of water weighs *1 kilogram.*

The container (top right) is *one cup of water.* This cup of water is approximately *14.44 cubic inches* and weighs approximately *0.52 pounds.*

Did you know that 1 cubic foot holds 7.481 gallons?

The use of the Metric System allows us to have a better understanding of the relationship between volume and weight. Remember . . . .

A **c**ubic **c**entimeter, which is approximately the size of a sugar cube, when filled with water, will weigh 1 gram and hold 1 milliliter. (Also called a **cc**.)

> 1 milliliter of water weighs one gram.

A cubic decimeter, which has sides measuring 1 *decimeter* or 10 centimeters each, when filled with water will weigh 1 kilogram and hold 1 liter.

> 1 liter of water weighs one kilogram.

The metric system is a much easier (NO FRACTIONS!), more efficient system than the one we currently use. Although we are more familiar with our own U.S. Standard Measures and a change would be difficult for most Americans, it has become necessary to understand the metric system.

Did you know that all countries in the world except the United States and Guam use the metric system? I am sure you have noticed that more and more products now come with dual labeling. Liters and milliliters are side by side with ounces, quarts, and gallons, while grams and kilograms are next to ounces and pounds.

CONVERSION USING UNITY FRACTIONS

There are many approaches to converting between units. Study the following examples to see one clever way to convert units. But first let me introduce you to:

1. **Unit:** The word *unit* has many meanings. It means *one* in some situations, or it can refer to a type of measurement. Pounds, ounces, feet, and miles are all types of *units*. You may even have heard someone say "I moved into the last *unit* in the complex."

2. **Unit Fraction:** A *unit fraction* is a fraction with a specific unit of measurement in the numerator or top of the fraction, and a 1 in the denominator, or bottom of the fraction. It is used as the starting fraction in a conversion. The numerator is the quantity that you want to convert.

$$\frac{6 \ gallons}{1}$$

3. **Unity Fraction:** A *unity fraction* is a fraction in which the top quantity equals the bottom quantity, while the numbers and units are different.

$$\frac{4 \ quarts}{1 \ gallon}$$

We all remember learning that $4 \times 1 = 4$, $6 \times 1 = 6$, and $0 \times 1 = 0$. We were told that the value of the original number does not change when multiplied by 1.

Similarly, when you studied fractions, you were told that:

$$\frac{3}{3} = 1 \text{ and if } 6 \times 1 = 6 \text{ then } 6 \times \frac{3}{3} = 6 \text{ as well.}$$

You thought about this and discovered that sure enough,

$$6 \times \frac{3}{3} = \frac{18}{3} = 6$$

Unlike numerical fractions like 3/3 or 7/7 that equal the value **1**, a unity fraction has different units in the numerator (on the top) than it does in the denominator (bottom).

$$\frac{4 \ quarts}{1 \ gallon} \quad \text{or} \quad \frac{2 \ pints}{1 \ quart}$$

Although you don't think 1 when you see a unity fraction, in a mathematical equation a unity fraction acts like the "multiplicative identity" 1. When you multiply by a unity fraction, you *do not change the value* of the original quantity, you just change its units!

Consider the fraction 3/4. Can you change its looks but not its value?

$$\frac{3}{4} \times \frac{6}{6} = \frac{18}{24}$$

Do 3/4 and 18/24 have the same value? Yes, they just look different.

Conversion with a unity fraction uses this same principal. You keep the value, but change the unit.

Suppose you know that the price of 3 pounds of lamb is $6.38. You would like to know the price of 9 ounces of lamb. You begin with a unit fraction whose numerator is the quantity you want to convert, then *multiply by 1* in the form of unity fractions.

$$\frac{9 \text{ ounces of lamb}}{1} \times \frac{\$6.38}{3 \text{ pounds of lamb}} \times \frac{1 \text{ pound of lamb}}{16 \text{ ounces of lamb}} = \$1.19625$$

Cancel using the principals of fractions . . .

$$\frac{9 \cancel{\text{ounces of lamb}}}{1} \times \frac{\$6.38}{3 \cancel{\text{pounds of lamb}}} \times \frac{1 \cancel{\text{pounds of lamb}}}{16 \cancel{\text{ounces of lamb}}} \cong \$1.20$$

Example 1: Convert 3 gallons to cups.

What do we know? We know there are 4 cups in 1 quart and 4 quarts in 1 gallon.

Step 1: Start with the unit fraction. $\dfrac{3 \text{ gallons}}{1}$

Step 2: Multiply by unity fractions to cancel unwanted units, while not changing the value of the original quantity.

$$\frac{3 \text{ gallons}}{1} \times \frac{4 \text{ quarts}}{1 \text{ gallon}} \times \frac{4 \text{ cups}}{1 \text{ quart}} = \frac{3 \times 4 \times 4 \text{ cups}}{1} = 48 \text{ cups}$$

Cancel gallons and quarts.

Step 3: 3 gallons are equal to 48 cups.

Example 2: Convert 350 milliliters to ounces.

What do we know? We know that 1 fluid ounce = 29.58 mℓ.

Step 1: Start with the unit fraction. $\dfrac{350 \text{ m}\ell}{1}$

Step 2:   Multiply by unity fractions to cancel out unwanted units.

$$\frac{350\ ml}{1} \times \frac{1\ fluid\ ounce}{29.58\ ml} = 11.83\ fluid\ ounces$$

Step 3:   350 milliliters is approximately 11.83 fluid ounces.

Example 3:   Convert one ounce into tablespoons.

We know there are 8 ounces in 1 cup and that 1 cup equals 16 tablespoons.

Step 1:   Start with the unit fraction. $\dfrac{1\ ounce}{1}$

Step 2:   Multiply by unity fractions to cancel out unwanted units.

$$\frac{1\ ounce}{1} \times \frac{1\ cup}{8\ ounces} \times \frac{16\ tablespoons}{1\ cup} = \frac{16\ tablespoons}{8} = 2\ tablespoons$$

Cancel ounces and cups.

Step 3:   1 ounce equals 2 tablespoons.

Example 4:   Try using this method to convert 24 miles into yards.
(*1 mile = 5,280 ft. and 1 yard = 3 feet*)

$$\frac{24\ miles}{1} \times \frac{5280\ feet}{1\ mile} \times \frac{1\ yard}{3\ feet} = 42{,}240\ yards$$

Cancel miles and feet.

$\Rightarrow$ 24 miles = 42,240 yards.

Example 5:   Convert 75 minutes to hours. Leave as a decimal answer.

$$\frac{75\ \cancel{minutes}}{1} \times \frac{1\ hour}{60\ \cancel{minutes}} = \frac{75}{60}\ hours = 1.25\ hours$$

Example 6:   Convert $3\frac{2}{3}$ minutes to seconds.

$$\frac{3\frac{2}{3}\ \cancel{minutes}}{1} \times \frac{60\ seconds}{1\ \cancel{minute}} = 3\frac{2}{3} \times 60 = 220\ seconds$$

$\Rightarrow 3\frac{2}{3}$ minutes = 220 seconds.

Example 7: Convert 144 ounces to quarts.

$$\frac{144 \text{ ounces}}{1} \times \frac{1 \text{ cup}}{8 \text{ ounces}} \times \frac{1 \text{ quart}}{4 \text{ cups}} = \frac{144}{32} \text{ quarts} = \frac{9}{2} \text{ quarts} = 4\frac{1}{2} \text{ quarts}$$

$\Rightarrow$ 144 ounces = 4 1/2 quarts.

Example 8: Convert 20 cups to gallons.

$$\frac{20 \text{ cups}}{1} \times \frac{1 \text{ quart}}{4 \text{ cups}} \times \frac{1 \text{ gallon}}{4 \text{ quarts}} = \frac{5 \text{ gallons}}{4} = 1\frac{1}{4} \text{ gallons}$$

$\Rightarrow$ 20 cups = 1 1/4 gallons.

Example 9: Convert 5½ pounds to ounces.

$$\frac{5\frac{1}{2} \text{ pounds}}{1} \times \frac{16 \text{ ounces}}{1 \text{ pound}} = 5\frac{1}{2} \times 16 \text{ ounces} = \frac{11}{2} \times \frac{16}{1} \text{ ounces} = 88 \text{ ounces}$$

$\Rightarrow$ 5 1/2 pounds = 88 ounces.

Example 10: Convert 6.8 kilograms to pounds and ounces.

First, convert 6.8 kg to lbs.

$$\frac{6.8 \text{ kilograms}}{1} \times \frac{2.2 \text{ pounds}}{1 \text{ kilogram}} = 14.96 \text{ pounds}$$

Next, keep the whole pounds and convert the fraction of a pound (0.96) into ounces.

$$\frac{0.96 \text{ pounds}}{1} \times \frac{16 \text{ ounces}}{1 \text{ pound}} = 0.96 \times 16 \text{ ounces} = 15.36 \text{ ounces}$$

$\Rightarrow$ 6.8 kilograms $\cong$ 14 pounds 15.36 ounces

If you are purchasing for a meal, you would order 15 pounds.

Example 11: Convert 3 gallons to liters.

$$\frac{3 \; gallons}{1} \times \frac{4 \; quarts}{1 \; gallon} \times \frac{1 \; liter}{1.057 \; quarts} = \frac{3 \times 4}{1.057} \; liters \cong 11.4 \; liters$$

⇒ 3 gallons ≅ 11.4 liters.

Or, using a conversion table:

$$3 \; gallons \times \frac{3.79 \; liters}{1 \; gallon} \cong 11.4 \; liters$$

As you can see, this #10 can of cranberry is labeled in both pounds and kilograms. It shows 7 lb 5 oz and 3.32 kg.

Example 12: Convert 7 pounds 5 ounces to kilograms. (Let 5/16 represent 5 ounces.)

$$\frac{7\frac{5}{16}}{1} \; lbs \times \frac{1 \; kg}{2.2 \; lbs} = 7\frac{5}{16} \times \frac{1 \; kg}{2.2} = \frac{117}{16} \times \frac{1}{2.2} \cong 3.32 \; kg$$

Cancel the pounds, convert to an improper fraction, multiply, and reduce.

Exercise Set 12

Convert each of the following quantities to the units indicated. Show all of your work.

1. 1 pint to ounces

2. 1 quart to cups

3. 1 quart to ounces

4. 1 gallon to pints

5. 1 gallon to cups

6. 1 gallon to ounces

7. 1 cup to teaspoons

8. 1 pint to tablespoons

9. 1 ounce to tablespoons

10. 1 ounce to teaspoons

11. 3 gallons to quarts

12. 4 quarts to pints

13. 5 pints to cups

14. 3 gallons to ounces

15. 2 quarts to ounces

16. 6 pints to ounces

17. 1 cup to tablespoons

18. 2½ gallons to quarts

19. quarts to cups

20. 1½ gallons to pints

21. 3¾ quarts to cups

22. 4 quarts to cups

23. 1 kilogram to grams

24. 5 kilograms to grams

25. 6 liters to milliliters

26. 3.5 liters to milliliters

27. 245 milliliters to ounces

28. 804 milliliters to ounces

29. 0.36 liters to milliliters

30. 0.5 liters to milliliters

31. 0.36 liters to ounces

32. 8.7 liters to ounces

33. 2 cups to milliliters

34. 1½ cups to milliliters

35. 4½ quarts to liters

36. 15 quarts to liters

37. 14 kilograms to pounds

39. 5 pounds to kilograms

38. 0.45 kilograms to pounds

40. 4¾ gallons to liters

| **A pint is a pound the world around.** |

Although we don't often speak of "pounds of water," baking recipes are always prepared using weights rather than volumes. If the formula requires 3 pounds of water, you would use 3 pints of water.

$$\frac{3\ pounds}{1} \times \frac{1\ pint}{1\ pound} = 3\ pints$$

Keep in mind that this is only true for liquid measures that are the consistency of water, but it is a reasonable approximation for most liquids used in cooking.

41. How many cups of water would you use if a certain recipe calls for 1 pound of water?

42. How many quarts of water are needed if the required amount is 4 pounds?

43. Convert 20 pounds of water to:

a. pints

b. quarts

c. gallons

d. liters

44. Convert 8 pounds of milk to:

a. pints

b. quarts

c. gallons

d. liters

Challenge Problems:

45. Approximately how many pounds would a 486-gallon waterbed weigh, if the vinyl bed itself weighs 25 pounds?

46. How many liters of water will the water bed in question 45 hold?

47. The average bathtub holds 250 gallons of water. How much would all of that water weigh? How many liters of water does this tub hold?

48. Convert 2 lbs 8 oz granulated sugar to kilograms.

49. Convert 1 lb 8 oz water to liters.

50. Convert 8 milliliters of saline solution to teaspoons.

51. Convert 22 milliliters of food coloring to tablespoons.

52. Convert 152 grams of butter to pounds.

53. Convert 875 grams of flour to pounds.

CHAPTER 13

Adding Weights and Measures

milliliters	mℓ	liters	ℓ
milligrams	mg	grams	g
teaspoon	tsp	tablespoon	tbsp
ounce	oz	cup	C
pint	p t	quart	q t
gallon	gal	pound	lb
cubic centimeter	cc	kilogram	kg

ADDING AND SUBTRACTING MIXED UNITS

When you add standard three-digit numbers as in ones, tens, and hundreds, you know the process. If you get 15 in the ones column, you carry the 1 to the top of the tens column leaving the 5 in the ones place in your answer. In subtracting sometimes you must borrow. If you try to subtract 6 from 5 in the ones column of a problem, you borrow one from the tens making the 5 a 15, then you can easily subtract the 6 from 15.

When you add or subtract mixed units—for example, pounds and ounces—it is necessary to combine the pounds and ounces separately.

Say you get 25 ounces in the ounce column. What should you carry over to the pound column? Or when you are subtracting and you must take 6 ounces from 5 ounces? How many ounces would you borrow from the pounds column? Let's look.

Example 1: Add the following units. You may convert as you go, or at the end of an addition problem.

$$
\begin{array}{r}
6 \text{ pounds } 15 \text{ ounces} \\
+ \ 2 \text{ pounds } \ \ 6 \text{ ounces} \\
\hline
8 \text{ pounds } 21 \text{ ounces}
\end{array}
$$

Convert 21 oz = 16 oz + 5 oz ⟹ 1 lb. 5 oz.
then add the pound to the 8 pounds.

⟹   9 pounds 5 ounces

Example 2: Subtract the following units being careful to convert as you work.

$$
\begin{array}{r}
9 \text{ pounds } \ \ \ \ 5 \text{ ounces} \\
- \ 8 \text{ pounds } \ \ 14 \text{ ounces}
\end{array}
$$

⟹

$$
\begin{array}{r}
\overset{8}{9} \text{ pounds } \overset{21}{5} \text{ ounces} \\
- \ 8 \text{ pounds } 14 \text{ ounces} \\
\hline
7 \text{ ounces}
\end{array}
$$

Borrow from the 9 lb, add 16 oz to 5 oz.

Example 3 : Add or subtract as indicated. Notice how much more simple this is using the metric system!

a)
$$
\begin{array}{r}
7 \ \ell \ \ \ 255 \text{ m}\ell \\
+ \ 35 \ \ell \ \ \ \ \ 15 \text{ m}\ell \\
\hline
42 \ \ell \ \ \ 270 \text{ m}\ell
\end{array}
$$

Because 1mℓ = .001ℓ.
Just add normally.

$$
\begin{array}{r}
7.255\ell \\
+ \ 35.015\ell \\
\hline
42.270\ell
\end{array}
$$

b)
$$
\begin{array}{r}
1 \text{ kg } 260 \text{ g} \\
- \ \ \ \ \ \ 420 \text{ g}
\end{array}
$$

⟹

$$
\begin{array}{r}
1260 \text{ g} \\
- \ \ \ \ 420 \text{ g} \\
\hline
840 \text{ g}
\end{array}
$$

Convert 1 kg = 1000 g and add it to the 260 g.

Example 4: Subtract

$$
\begin{array}{r}
8\ell \ \ \ \ \ 6 \text{ m}\ell \\
- \ 3 \ \ell \ \ 150 \text{ m}\ell
\end{array}
$$

⟹

$$
\begin{array}{r}
\overset{7}{8} \ \ell \ \ \overset{1006 \text{ m}\ell}{6 \text{ m}\ell} \\
- \ 3 \ \ell \ \ 150 \text{ m}\ell
\end{array}
$$

⟹

$$
\begin{array}{r}
7 \ \ell \ 1006 \text{ m}\ell \\
3 \ \ell \ \ \ 150 \text{ m}\ell \\
\hline
4 \ \ell \ \ \ 856 \text{ m}\ell \text{ or } 4.856\ell
\end{array}
$$

or because 1 mℓ = 0.001ℓ, we can just rewrite the problem:

$$
\begin{array}{r}
8\ell \ \ \ \ 6 \text{ m}\ell \\
- \ 3\ell \ 150 \text{ m}\ell
\end{array}
$$

⟹

$$
\begin{array}{r}
8.006\ell \\
- \ 3.150\ell \\
\hline
4.856 \ \ell
\end{array}
$$

Example 5: Add.

```
                                     1
  3 g     650 mg            3 g     650 mg     Convert the 1000 mg to 1 g and carry it over to the g column.
  4 g     820 mg    ⟹    + 4 g     820 mg
  7 g    1470 mg            8 g    1̶4̶70 mg
```

or   3.650 g
                4.820 g   This is the metric system at it's finest!
                8.470 g

TO CONCLUDE

To add or subtract metric units, you just insert a decimal point after the *whole* gram or liter and place the milligrams or milliliters in the thousandth place. The metric system uses base 10, like our own decimal system.

$$0.001\ell = 1\ m\ell \quad\quad \text{or} \quad\quad \frac{1}{1000}\ \ell = 1\ m\ell$$

$$1000\ mg = 1g \quad \text{and} \quad\quad 1000g = 1\ kg$$

See how easy it is to add and subtract mixed units using the metric system!

```
  3 kg   52 g    ⟹     3.052 kg
 +7 kg  104 g          7.104 kg
                      10.156 kg
```

```
                                                  7
  7ℓ   181 mℓ    ⟹    7.181 ℓ    ⟹    7.18¹1
 −3ℓ     4 mℓ         −3.004 ℓ         −3.004
                                        4.177
```

Exercise Set 13

Add or subtract as indicated. Express your final answer in the largest whole units.

1.   3 lb 6 oz
     + 1 lb 8 oz

2.   8 lb 10 oz
     + 2 lb 8 oz

3.   7 lb 15 oz
     +   43 oz

4.   2 lb 4 oz
     + 5 lb 31 oz

5.   19 lb 13 oz
     +   64 oz

6.   1 lb 9 oz
     + 21 lb 25 oz

7.   6 lb 13 oz
     − 3 lb 14 oz

8.   9 lb 5 oz
     − 8 lb 14 oz

9.   5 lb
     − 3 lb 6 oz

10.   4 lb 7 oz
      − 3 lb 12 oz

11.   7 lb 5 oz
      − 3 lb 18 oz

12.   3 lb 1 oz
      − 1 lb 4 oz

13.   1 gal 5 qt 3 pt 1 c
      + 2 gal 2 qt 5 pt 6 c

14.   2 gal 2 qt   3 c
      +   7 qt 2 pt 1 c

15.   3 gal 7 qt 9 pt 8 c
      +   3 qt   3 c

16.   3 gal 1 qt 2 c
      −   3 qt 1 c

17.   4 gal 1 qt 3 pt 3 c
      − 1 gal 3 qt 1 pt 4 c

For questions 18–22, rewrite the quantities using a decimal point before adding or subtracting.

18.
```
      2 ℓ     256 mℓ
  +   3 ℓ      58 mℓ
```

19.
```
     14ℓ       25 mℓ
  −             330 mℓ
```

20.
```
    332 g     456 mg
  +  55 g      44 mg
```

21.
```
    726 kg     25 g
  −  33 kg    498 g
```

22.   3 ℓ 326 mℓ − 525 mℓ

23.   2 LB 8 oz + ½ oz + 3½ oz

⇓ Put your answers in the best unit.

24.   2¼ c + 1¾ c − 7/8 c

25.   5⅛ c + 3 pt

26.   2 qt + 7 c − 3 pt

27.   3⅛ lb. − ⅜ lb.

28.   ⅝ gal − 3 pt

Challenge Problems

29.   You purchased 40 pounds of top sirloin steak of which you sold 18 kg. How many ounces do you have left over?

30.   You chopped 8¾ ounces of onions but used 3 tablespoons. How many teaspoons do you have left?

31.   If you take 6 liters from a 5-gallon jug, how many ounces do you have left?

32.   You added 13.63 liters of coolant to your new radiator. It holds 5 gallons. How many more cups of coolant do you need to add to your radiator?

33.   Would 6¾ pounds of water would fit into a gallon pitcher? If yes, how many more fluid ounces would the pitcher hold? If no, how many fluid ounces of water would you have left over?

34.   The doctor ordered you to take 15 cc of cough syrup every four hours. Write this amount in three additional units. (label)

35.   One liter of water weighs 1 kilogram. Approximately how much does 1 quart of water weigh?

CHAPTER 14

Costing

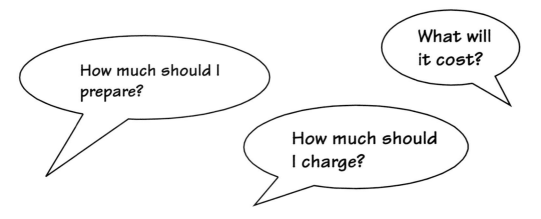

A very important factor when figuring the amount to prepare is the yield. In Chapter 9 yield was used to describe the total packaged amount of a given food item on a purchase order. Here the word *yield* is used to specify *the amount of usable food left after cooking, boning, and trimming.* After you trim and cook a roast that cost $1.15 per pound, will it still be worth $1.15 per pound? No.

You will trim portions of most foods such as fruits, vegetables, meat, poultry, and fish before cooking them, causing the weight of the purchased food to go down. Moreover, during the cooking process, foods often shrink and lose more weight.

What is the approximate usable yield after cooking a 10-pound leg of lamb? Perhaps only 5 pounds of usable meat remain. Or maybe 8 pounds. These are questions you will have to be prepared to answer before you can figure the cost of any dish you prepare.

Oftentimes, it is not obvious how much the food you sell actually costs you. Think about buying a 22-pound *Chinook* or king salmon. Suppose you pay the fisherman $2.80 per pound, or $61.60 for the

King (Chinook) is the largest of the salmon, averaging 10 to 40 pounds. Most king have red flesh although some white kings are very light colored.

whole fish. How much of the fish will you be able to use? After trimming, cleaning, and cooking, what would the price per (usable) pound be?

If you are a Yupik Eskimo, part of the salmon will be used for steaks, part for fish-head soup, the reddest meat mixed with rock salt to cure, some of the roe mixed with baking powder to be used for fish bait, and the remainder, guts and all, will be mixed with rice to feed to your dog team. In this case all of the salmon will be used for food.

Unfortunately, however, most food service establishments are not able to use every part of a fish or a side of beef, pork, or lamb. When figuring cost, you need to consider shrinkage from cooking and waste from boning and trimming.

One establishment uses the following worksheet to determine food cost.

FOOD YIELD AND COST WORKSHEET

FOOD ITEM	PURCHASE PRICE UNIT MEASURE	COST PER POUND	CUT & COOK YIELD%	COST PER USABLE POUND	PORTION SIZE	COST PER PORTION
LETTUCE	$21.24–20 pounds	$1.062	80%	$1.327	8 oz	$0.664
BLEU CHZ	$43.80–18 gal	$2.43 gal	100%	$0.189 oz	2 oz	$0.038
PORK CHOPS	$48.20–35/5 oz	$1.377 ea	76%	$1.377 each	2 chops	$2.754
RIBEYE	$79.85–10 pounds	$7.985	91%	$8.774	6 oz	$3.290
AHI FILLETS	$390.00–30 pounds	$13.00	80%	$16.250	7 oz	$7.109
WHOLE AHI	$240.00–30 pounds	$8.00	56%	$14.285	4 oz	$3.571
SALMON	$45.66-10 pounds	$4.566	50%	$9.132	8 oz	$4.566

Note that for the pork chops, although 24% is lost through cut and cooking, this does not affect the price. It does, however, change the portion from two 5-ounce chops to two 3.8-ounce chops.

Take a look at the first line. Lettuce begins at $21.40 for 20 pounds.

$$\$21.24 \div 20 \cong \$1.062 \text{ per pound, purchase price}$$

After cleaning and removing bad leaves, the yield is determined to be only 80% of the original head of lettuce. The 20 pounds of lettuce is now (80% of 20 pounds) 16 pounds of lettuce through cleaning and trimming.

You conclude that you bought 20 lb of lettuce for $21.24; however, after cleaning you have only 16 pounds of lettuce. You then find the price of the useable lettuce.

$$\$21.24 \div 16 \cong \$1.327 \text{ per usable pound.}$$

The cost per 8 oz portion then becomes ½ lb at $1.327 per pound or $\cong$ $0.664 per serving.

Let's look at Ahi Tuna, lines 5 and 6. It appears that buying the whole fish will cost considerably less per pound than buying fillets. However, the yield is shown as 56% (depending on your use). This

would mean that for a 30 lb Ahi, you may be able to use only 16.8 pounds, and the price per *usable pound* would be:

$$\$240.00 \div 16.8 \text{ lbs.} \cong \$14.286 \text{ per useable pound.}$$

If you are lucky enough to find a few 2½ oz servings of sashimi and cut a few 8 oz fillets before making the rest into Ahi Poke, you may make a nice profit on your fish, although you had to do the trimming and cleaning yourself.

Perhaps you sell only Ahi steaks on your menu. In this case you may want to purchase the fillets. Let's assume that the usable portion is approximately 80% and the price of the #2 Med Ahi Block that you choose is $13.00 per pound. You purchase 30 pounds.

$$30 \text{ lbs. @ } \$13.00 \text{ per/lb.} = \$390.00$$

You are able to use 80% of the 30 pounds, 0.80 x 30, or 24 pounds. With the purchase price of $390, you get 24 pounds of usable fillets. This translates to:

$$\$390.00 \div 24 \text{ lbs.} = \$16.25 \text{ per pound} \cong \$1.02 \text{ per ounce}$$

Knowing the *yield* of any given food that you purchase is a *MUST*. It is with the knowledge of exactly how much the food you are serving costs *you* that you are able to determine menu prices.

Let's sidetrack for a moment. Suppose you serve lobster tail on your menu. Tails that are sized 7–8 oz might sell for $18/lb, while tails sized 9–10 oz are going for $15/lb. You want to decide which is a better deal.

Compare the price per ounce. The smaller tails are $18 ÷ 16 oz or $1.13 per oz, while the larger tails are $15 ÷ 16 oz or $0.94 per oz. From this information, you may think that the larger tails are "cheaper." However, if you choose the larger tail for 94¢ per ounce, one 10 oz tail will cost $9.40. If you choose the smaller tail, you will pay $1.13 × 8 or $9.04 for an 8 oz tail. Thus the smaller tails cost approximately 36¢ less than the larger tails.

PORTION SIZE AND COST PER SERVING

Let's assume for a moment that you manage a large restaurant, and you happen to walk by a new employee preparing a dinner plate that is about to be served. You observe that the portion of mashed potatoes looks exceptionally large. You tell him that the serving is too large and that he needs to be consistent in his serving size.

"Why?" he answers, "We have plenty of mashed potatoes prepared, and it is almost closing."

What would you reply? Why does portion size matter? Does consistency matter?

The bottom line here is profit. To make a profit on each plate that you serve, you need to be aware of the exact cost of the food placed on each plate. If you can determine the cost of the food used per plate, you will be able to charge accordingly on your menu.

Realistically, unless you measure and weigh exactly each portion of food served, there will always be a bit of inconsistency in portion size. To improve consistency, many fast food establishments use a scale to weigh each ounce of ice cream put into a milk shake or use only a number 16 scoop leveled with meat filling for each taco.

In any case, you must *know* how much money you are spending to figure out how much you are making.

> To figure the cost per serving, divide the total cost of the food by the number of servings it yields.

Example 1: Find the cost of a 4-ounce serving of stewed tomatoes, given that the cost of six #10 cans of stewed tomatoes is $23.80. The total yield of all six cans is approximately 636 ounces.

Step 1: First find the number of servings of stewed tomatoes yielded.

636 ounces ÷ 4 ounce servings = 159 servings

Step 2: Now divide the food cost by the number of servings.

$23.80 ÷ 159 ≅ $0.149

Step 3: A 4 oz serving of stewed tomatoes cost approximately 15¢.

Example 2: How many 4-ounce servings of stewed tomatoes can you get from one #10 can? Approximate can yield 106 ounces.

This is simply a division problem. You *divide the total yield of the can by the serving size* to find the number of servings of a given size.

106 oz ÷ 4 oz = 26.5 servings

Example 3: Onion rings come in cases that contain eight 2 1/2-pound boxes and yield 800 pieces per case. How many cases of onion rings need to be ordered to feed 450 hungry kids? Assume 8 pieces each child.

Step 1: Since there are 800 pieces per case, and each child gets 8 pieces there are:

$$800 \text{ pieces} \div 8 \text{ per child} = 100 \text{ servings per case}$$

Step 2: You need 450 servings and there are 100 servings per case, so you will need:

$$450 \text{ servings} \div 100 \text{ per case} \Rightarrow 4\frac{1}{2} \text{ cases are necessary}$$

Step 2: Since you cannot order parts of a case, you must order 5 cases of onion rings.

Example 4: Referring to Example 3, how many 5-ounce servings of onion rings will you have left over?

Step 1: You used $4\frac{1}{2}$ cases for the kids, which leaves you $\frac{1}{2}$ case.

Step 2: Each case contains 8/2.5lb boxes or 20 pounds of onion rings.

Step 3: Each $\frac{1}{2}$ case contains 10 pounds or 160 ounces of onion rings.

Step 4: Divide the total ounces by the serving size to find the number of servings.

$$160 \text{ oz} \div 5 \text{ oz servings} = 32 \text{ servings}$$

$$\Rightarrow \text{There will be 32 servings left over.}$$

Example 5: Approximately how much does one onion ring from Example 2 weigh and how much does each ring cost, given that one case sells for $35.09?

Step 1: Find the total weight of one case of onion rings. You know that one case contains eight $2\frac{1}{2}$ pound boxes. The total weight:

$$8 \text{ boxes x } 2\frac{1}{2} \text{ lb.} = 20 \text{ pounds}$$

Step 2: Divide the total weight by the total number of pieces to find weight per piece.

$$20 \text{ pounds} \div 800 \text{ pieces} = 0.025 \text{ pounds per piece}$$

$$or \quad 0.025 \text{ x } 16 = 0.4 \text{ ounces per piece}$$

Step 3: The approximate cost of one piece will be the total cost of the case divided by the total number of pieces.

$$\$35.09 \div 800 \text{ pieces} \cong \$0.0439 \quad \text{or } 4.4\text{¢ per piece.}$$

Step 4: The weight of one onion ring is approximately four-tenths of one ounce. The approximate cost of one onion ring piece is 4.4¢.

FOOD COST PERCENT

After you are comfortable with figuring out exactly how much each serving of any given food costs *you*, then you are ready to decide how much to charge your customer. Many free-standing restaurants try to keep *their* cost of raw food used in preparation of a dish to **no more than 35%** of what they charge customers for that menu item.

This concept is called *food cost percent*. The phrase is used to describe the *ratio of Your Monthly Food Purchase : Monthly Food Sales* in terms of a percent. (This is another example of the concept you learned in Chapter 5, comparing the part to the whole. You may want to look back and reread it.)

Your monthly food purchases are your *Raw Food Cost*. This is the amount of money you spend buying the foods that make up the menu items you sell, which then become your monthly food sales!

Extremely high volume businesses may spend only 20% of their monthly sales on food purchases, while a service-type institution, such as a school cafeteria, may go as high as a 75% food cost.

Example 6: Your monthly food purchases for October were $25,469, while your monthly food sales was $66,900. Find your food cost percent for October.

Step 1: Write the ratio as a fraction.

$$\frac{Monthly\ Food\ Cost}{Monthly\ Sales}$$

Step 2: Calculate:

$$25,469 \div 85,840 \cong 0.2967$$

Step 3: State:

The food cost for October was 29.7%.

MARKUP AND MENU PRICING

Let's say you decide to maintain a *33% food cost ratio*. All you need to do is a little division to find the correct prices for your menu. Begin with a word sentence that describes what you want, and then translate it into a mathematical equation.

33% of what I charge my customers for menu items, will go directly
to pay for the raw food used in the preparation of those menu items.

33% of the Menu Price = Raw Food Cost

Menu Price = Raw Food Cost ÷ 33%

Menu Price = Raw Food Cost ÷ 0.33

> **Algebra Review**
> To solve for menu price, divide both sides of the equation by 33%.

It follows from the above example that to maintain a 33% food cost, simply divide your raw food costs by 0.33 to get your menu price.

Mark up formula (division):

Menu Price = Raw Food Cost ÷ Food Cost Percent

Example 7: Find the selling or menu price for the following foods, given that you want your food cost percent to be 21. *Use the division method.*

a.   Baked Potato, your cost $0.187 each.
b.   Peach Slices, your cost $0.079 per ounce.
c.   6 ounce Hot Chicken Wings, your cost $0.256 per ounce.
d.   Boneless/Skinless Chicken Breast $1.89 each, your cost.

Your Cost ÷ Food Cost %		Menu Price
a) 0.187 ÷ 0.21	$\Rightarrow$	$\cong 0.8904 \cong$ 89¢ potato
b) 0.079 ÷ 0.21	$\Rightarrow$	$\cong 0.3762 \cong$ 38¢ peach slices
c) 0.256 ÷ 0.21	$\Rightarrow$	$\cong 1.2190$ per oz
		$\cong$ \$1.22 x 6 oz = \$ 7.32 hot wings
d) 1.89 ÷ 0.21	$\Rightarrow$	$\cong$ \$9.00 chicken

Some people prefer a multiplication format to derive menu prices. Apply your knowledge of fractions, and the fact that $33\% = 0.33 = \frac{33}{100}$. Then begin once again.

$$33\% \text{ of the Menu Price} = \text{Raw Food Cost}$$

$$\text{Menu Price} = \text{Raw Food Cost} \div 33\%$$

$$\text{Menu Price} = \text{Raw Food Cost} \div \frac{33}{100}$$

$$\text{Menu Price} = \text{Raw Food Cost} \times \frac{100}{33}$$

$$\text{Menu Price} = \text{Raw Food Cost} \times 3.\overline{03}$$

Mark up formula (multiplication):

$$\text{Menu Price} = \text{Raw Food Cost} \times \frac{100}{\text{Food Cost}}$$

The *100 ÷ Food Cost* is called the *markup* and can be used to determine the selling price of menu items. One of these two methods can be used when you want to maintain a specific food cost percent.

Here are the **steps to finding a menu price**, using the multiplication method.

Step 1:   Decide your necessary food cost percent.

Step 2:   Divide 100% by the food cost % to determine a markup.

Step 3:   Multiply your raw food cost by the markup to get your menu price.

Example 8:   How much should you charge the school for the 450 servings of onion rings, if you want a 30% food cost? *Use the multiplication method.*

Step 1:   Remember that the kids ate 4½ cases at $35.09 per case.

$$\text{Raw Food Cost} \cong 4\tfrac{1}{2} \times 35.09$$
$$\text{Raw Food Cost} \cong 157.905$$
$$\text{Raw Food Cost} \cong \$157.91$$

Step 2:   Find the markup using a 30% food cost.

$$100\% \div 30\% \Rightarrow 100 \div 30 \quad \textit{Canceling the \% signs.}$$
$$\cong 3.33$$

Step 3:   Using the markup, find the selling price.

$$\text{Selling Price} \cong \text{Raw Food Cost} \times \text{Markup}$$
$$\text{Selling Price} \cong 157.91 \times 3.33$$
$$\text{Selling Price} \cong 525.840$$

Step 4:   The school should be charged $525.84 for the onion rings.

> And how much is that per child?   Hmmm...
> $525.84 ÷ 450 = $1.17

Example 9: Find the markup and menu price for the following, using both the multiplication and division methods. (*do not round* until your final answer)

	Raw Food Cost	Food Cost	Markup	Menu Price
a.	14.60	38%	?	?
b.	89¢	29%	?	?
c.	3.00	33%	?	?

a) Find the markup: $100 \div 38 = 2.632$

     Menu Price - *Multiplication* $14.60 \times 2.632 = 38.427 \cong \$38.43$

     Menu Price - *Division* $14.60 \div 0.38 = 38.421 \cong \$38.42$

b) Find the markup: $100 \div 29 = 3.448$

     Menu Price - *Multiplication* $0.89 \times 3.448 = 3.069 \cong \$3.07$

     Menu Price - *Division* $0.89 \div 0.29 = 3.069 \cong \$3.07$

c) Find the markup: $100 \div 33 = 3.030$

     Menu Price - *Multiplication* $14.60 \times 3.03 = 44.238 \cong \$44.24$

     Menu Price - *Division* $14.60 \div 0.33 = 44.242 \cong \$44.24$

Sometimes you will want to compare your food cost markup with other food service establishments. Here is a formula that will make it easy to find a markup when you know a menu price and the raw food cost.

To find the markup when you know the selling price and the raw food cost, divide.

Menu Price ÷ Raw Food Cost = Markup
$44.24 \div 14.60 \cong \text{Markup}$
$3.030 \cong \text{Markup}$
$3.03 \cong \text{Markup}$

Today companies list product prices in exact weight more often than in can sizes. However, some distributors still use can numbers in conjunction with the exact measure. The number symbol # is used *before* the actual number to identify can size, as in a #10 can, and *after* the number, as in 10#, to represent pounds.

Keep in mind that although two different foods may be packed in the same size container, their weights may differ quite a bit. For instance a #10 can of Cranberry Sauce weighs 7 lb 5oz, while a #10 can of hot fudge weighs 6 lb 12 oz.

6/#10	NIFDA TOMATO STEWED	636 OZ	$19.38
6/#10	APPLESAUCE MICHIGAN	648 OZ	$12.04
24/#303	APPLESAUCE WISCONSIN	384 OZ	$15.05
4/54oz	CREAM OF BROCCOLI BASE		$32.20
200/12gram	BBQ SAUCE PACKETS		$7.85
6/1qt	GARLIC CHOPPED		$4.40
4/4.25#	BEEF STEW		$12.20
4/5#	BLEU CHEESE CRUMBLES		$49.76
6/#10	CHILI SAUCE	690 OZ	$18.76
24/#2.5	PEACH SLICES	418 OZ	$24.58
6/.5 gal	STRAWBERRY TOPPING		$ 8.05
6/#5	HOT FUDGE SAUCE	48 OZ	$ 3.41
4/5#	GOUR WHOLE COFFEE BEAN		$83.65
CS/8-3#	COFFEE BLENDERS		$10.05
CS/360-3/8 oz	ULTRA CREAMERS		$12.65
12/2#	EGG BEATERS		$42.05
15/DOZ	EXTRA LARGE EGGS GRADE A		$ 8.20

Price list for Exercise Set 14.

Exercise Set 14

For questions 1–12, fill in the following Food Yield and Cost Worksheet.

FOOD YIELD AND COST WORKSHEET

FOOD ITEM	PURCHASE PRICE UNIT MEASURE	COST PER POUND	CUT & COOK YIELD%	COST PER USABLE POUND	PORTION SIZE	COST PER PORTION
LETTUCE	$58.76–50 pounds		75%		8 oz	
ALMONDS	$28.63–5 pounds		100%		2 oz	
AHI STEAK	$48.20–35/5 oz	per fillet	80%	per fillet	2 fillets	
RIBEYE	79.85–10 pounds		91%		6 oz	

Use the price list on the previous page to answer questions 13–28.

13.   Find the cost of an 8-ounce serving of applesauce from Michigan.

14.   Find the cost of an 8-ounce serving of applesauce from Wisconsin.

15.   Find the cost of 2 ounces of cream of broccoli soup base.

16.   How much would 1 cup of cream of broccoli soup base cost?

17.   What would the price of 85 packets of BBQ sauce be?

18.   What would 1,000 packets of BBQ sauce cost?

19.   If you ordered one case of Ultra Creamers and used only 12 dozen packets, how many packets of creamers would you have left?

20. How much Coffee Blenders would you have left from a case if you used 13 pounds?

21. What is the price of 2 dozen extra large grade A eggs?

22. How much would 3 ounces of egg beaters be?

23. One pound of Gourmet Coffee Beans would cost how much?

24. Sixty pounds of Beef stew would cost how much?

25. A dish of peach slices weighs in at 7 ounces. How much is it worth?

26. How much would 3 gallons of Strawberry Topping cost?

27. If you used 1 ounce of Strawberry Topping for a malt, how much did the topping cost?

28. Find the cost of a 2 ½-ounce serving of a peas and carrots mix if a 5-pound box of peas costs $5.15 and a 2 ¼ box of carrots costs $3.03.

For questions 29–33 Find the *food cost percent* and the *markup* that would be used to maintain that specific food cost percent.

	Monthly Food Cost	Monthly Sales	F.C. %	Markup
29.	$ 18,356	$ 48,305	_____	_____
30.	$ 6,407	$ 20,668	_____	_____
31.	$ 618	$ 1,507	_____	_____
32.	$ 4,085	$14,590	_____	_____
33.	$ 9,735	$ 33,000	_____	_____

34. 1# - Honey Smoked Turkey Breast $3.50
What should you charge for an 8 oz portion, if your food cost percent is 29?

35. 48/4oz - Turkey Burgers $18.00
What should you charge for one burger if you are working at a 36% food cost?
Keep in mind that this is just the meat, not the price of all ingredients going into the burger!

36. 20# - 9" Polish Sausage $23.85
How much would you charge for 1 polish sausage that weighs 10 oz if you want to maintain a 31% food cost?

37. 15# - Pure Beef Patty Thick and Juicy $22.85
What would you charge for a 1/4 lb burger, keeping your food cost percent to 38? *Keep in mind that this is just the meat, not the price of all ingredients going into the burger!*

38. The selling price of one 12″ pepperoni pizza is $14.60. If your raw food cost was $3.00, what was your markup?

39. The menu price of a 9″ cheese and sausage pizza is $8.50. Your raw food cost is $1.75. What was the markup?

40. 20/8oz – Ribeye $ 80.85
    27/6oz – Ribeye $ 80.85

    a. Using a markup of 2.69, find the menu price for one 6 oz ribeye steak.

    b. Find the menu price for one 6 oz ribeye, given a 33% food cost.

    c. Find the markup if the menu price for one 8 oz ribeye steak is $16.00.

    d. Which steak is less expensive for you to buy? The 8 oz or the 6 oz?

41. CS 5 oz – Spencer Steak 32ct $ 31.50
    CS 8 oz – Spencer Steak 20ct $ 28.85

    a. Find the price of one 5 oz steak.

    b. Find the price of one 8 oz steak.

    c. Using a 3.88% markup, find the menu price of a 5 oz steak.

d. Using a 4.21 markup, find the menu price of an 8 oz steak.

e. Find the markup and the menu price of a 5 oz steak given a 31% food cost.

f. Find the menu price of an 8 oz steak, given a 31% food cost.

42. 6# Sloppy Joe Filling $ 9.65
     5# Au Gratin Potatoes $ 8.05
     8# Chicken Pot Pie Filling $12.25

Find the markup and the menu price for 6 ounces of each item above, given the food costs as:

		Markup	Menu Price	
a.	41%	_____	_____	Sloppy Joe
b.	27%	_____	_____	Potatoes
c.	35%	_____	_____	Pot Pie

Challenge Problems

Use the price list given for these exercises when needed.

43. Which product is cheaper?

     6/#10 HOT FUDGE TOPPING $7.35   *Assume #10 can ≅ 13 cups.*
     6/#5   HOT FUDGE TOPPING $ 3.41 *Assume # 5 can ≅ 6 cups.*

44. How much would 1 liter of Cream of Broccoli Soup Base cost?

45. Find the cost of a 6-ounce glass of wine given the cost at $35.60 per 18 ℓ box.

46. Find the cost of 100 grams of BBQ sauce.

47. Find the cost of 6 ounces of BBQ sauce.

48. Find the price of 650 packets of BBQ sauce.

49. How much would 10 ounces of Ultra Creamer cost?

50. How much would a mixture of 1 quart of chili sauce, 3 oz chopped garlic, 5 eggs, and 4 ounces of bleu cheese crumbles cost?

51. Find the total yield in milliliters, of six #2 cans of Pineapple Juice if each #2 can holds approximately 1 pt 2 oz.

52. A number 10 can is equivalent to approximately 7 #303 cans. If one #10 can of cranberry sauce weighs 7 lb 5 oz, how much does one #303 can of Cranberry Sauce weigh in kg?

53. Twenty-four #303 cans hold approximately 384 ounces. How many milliliters does one #303 can hold?

CHAPTER 15

Recipes: Yields, Converting, Costing

Fresh Créme Glacé

Ingredients: 950g fresh cream, 50g orange powder, 100g sugar powder, 15g agar, 20g cocoa, 10g red cherries, a dash of green edible pigment.

Making:
1. Melt the agar, mix with orange powder and sugar powder, evenly for use.
2. Beat the fresh cream till risen, pour the mixture of agar, sugar powder, and orange powder, then place in the glasses.
3. Add a dash of green pigment into little amount of fresh cream, stir into green color, make cream shreds covering the whole surface of the glass, make cream pandas on the green shreds, then garnish with red cherries.

How much fresh *Cream Glass* does the above recipe from China yield? What is the bulk yield? What is the portion yield?

To answer these questions, look at the picture on the next page. The picture suggests a six-serving yield. Can you guess the portion size (in grams) from the picture? Try to estimate the bulk yield as well.

The **recipe yield**, or the amount of prepared food you get from a recipe, is the first thing on a recipe card that a chef must look at when deciding how much to prepare. The yield of any recipe can be described in two ways. The **bulk yield** is used to describe the total amount of food that will be prepared if the recipe is followed correctly. Estimating the bulk yield for the Cream Glass recipe above, I'd say the bulk yield is about 1000 grams or one kilogram. The **portion yield** tells the number of servings and their sizes. The portion yield would be approximately 1000 grams ÷ 6 servings = 167 grams each.

What would happen to the above six servings if you decided that the portion size was too small? Naturally, there would be fewer than six servings if you increased the serving size. It is important to know how big you want your portions to be and how many portions you will need before you prepare any recipe.

EXAMPLE 1: The portion yield for a certain recipe of chocolate swirl cheesecake is listed as eight 3 oz servings. What is the bulk yield?

Step 1: Multiply to find total ounces.   8 x 3 oz = 24 oz

Step 2: Convert to best measure.

$$\frac{24\ oz}{1} \times \frac{1\ lb}{16\ oz} = \frac{24\ lb}{16} = 1\frac{1}{2}\ lb$$

Step 3: The approximate bulk yield for the recipe is 1½ pounds.

EXAMPLE 2: The bulk yield of a muffin recipe is 15 pounds 3½ ounces. Describe the portion yield if each muffin is to have 2 ounces of batter.

Step 1: Convert bulk yield into ounces.

$$\frac{15\ lb}{1} \times \frac{16\ oz}{1\ lb} = 240\ oz + 3\frac{1}{2}\ oz$$

Step 2: Divide total ounces by portion size.   243.5 oz ÷ 2 oz = 121.75 servings

You must decide whether to "short" each muffin to make 122 muffins, or to have ¾ of a muffin left over. You are only short ¼ of a 2 ounce muffin or ½ ounce short. If you divide the ½ ounce among 122 muffins, it will <u>not</u> be noticeable.

Step 3: The portion yield would read 122 2-ounce muffins.

CONVERTING RECIPES

Converting recipes is merely a way of making any recipe fit your need. If, for instance, you need 60 servings of mashed potatoes and the recipe that you want to follow has a listed yield of 20 servings, you naturally multiply each ingredient amount by 3 to produce the correct yield.

Some conversions are not quite as simple as others, so you may have to do a bit of arithmetic to figure the new ingredient amounts. Take, for example, a recipe that has a listed yield of 15 servings. You intend to make enough for 23 servings. How would you go about converting the ingredient amounts?

Before I give you a basic working formula for converting recipes, let me draw on the knowledge of proportion that you already possess. This way if you forget the formula that I am about to illustrate, you can always use the ratio and proportion method.

EXAMPLE 3: You need 2½ cups of sour cream to make a special cream sauce for a beef tenderloin dish that yields 15 servings. How much sour cream would you need for the sauce for 23 servings?

Step 1: Using ratio and proportion:
$$\frac{sour\ cream}{serving} = \frac{2.5\ cups}{15} = \frac{?}{23}$$

Step 2: Cross multiply:
$$2.5\ cups \times 23 = 15 \times ?$$
$$57.5\ cups = 15 \times ?$$

Divide:
$$57.5 \div 15\ cups = ?$$
$$3.833\ cups = ?$$

Step 3: You will need 3.8 cups of sour cream to make 23 servings of cream sauce.

$$3.8\ cups = 3\ cups + \left(\frac{0.8c}{1} \times \frac{8oz}{1c} \right) \cong 3\ cups\ 6^{1/2}\ ounces.$$

Using ratio and proportion will always work. However, a single conversion factor is a bit faster. To find your **conversion factor** or **C.F.** (similar to your markup rate) you simply use the ratio of the new yield to the old yield. Using the previous example:

$$C.F. = \frac{New\ Yield}{Old\ Yield} = \frac{23\ servings}{15\ servings} \cong 1.533$$

Next, just *MULTIPLY THE OLD RECIPE AMOUNT BY THE CONVERSION FACTOR* to find your new recipe amount.

$$2½\ cups\ of\ sour\ cream \times 1.533 \cong 3.83\ cups\ of\ sour\ cream \cong 3c\ 6½\ oz$$

EXAMPLE 4: You are preparing Layered Salmon Pie for 72 people. You plan to feed each person 1/4 pie. Convert the recipe to the necessary amounts.

Layered Salmon Pie

Ingredients: Yield 30 pies

7.5 lb. Salt pork
30 lb. Salmon steak; or fillets skinned & cut into pieces
3¾ c Flour; all purpose
7½ c Celery; chopped
15 oz Onion; finely chopped
30 Potatoes; medium, peeled & sliced
Salt & ground white pepper to taste
30 - Pastry for double crust 9" pie

Step 1: Decide how many pies you need to prepare.

Each pie yields 4 servings, and you need 72 servings. Thus you divide to find the amount of pies needed.

72 servings ÷ 4 servings per pie = 18 pies needed

Step 2: Find the conversion factor for the recipe.

$$\frac{new\ amount}{old\ amount} = \frac{18\ servings}{30\ servings} = 18 \div 30 = 0.6$$

Step 3: Find each new ingredient amount by multiplying by the conversion factor.

Ingredients for 30 pies	Multiply by C.F.	Ingredients for 18 pies
7.5 lb. Salt pork	7.5 × 0.6	4.5 lb. Salt pork
30 lb. Salmon steak	30 × 0.6	18 lb. Salmon steak
3 ¾ c Flour	3.75 × 0.6	2¼ c Flour
7.5 c Celery	7.5 × 0.6	4½ c Celery
15 oz Onion	15 × 0.6	9 oz Onion
30 Potatoes	30 × 0.6	18 Potatoes
30 Pastries	30 × 0.6	18 Pastries

RECIPE COSTING

Costing a given recipe simply means finding out how much money you spent on ingredients to create that recipe. Sometimes it may be very straightforward, other times not so straightforward. For example, let's say that you need 7 pounds of cottage cheese for a particular recipe. The distributor lists the price of cottage cheese as *5# FOR $6.30.* You have learned that you need only find the price per pound, then multiply by 7 to get your ingredient cost.

$6.30 ÷ 5 lb. = $1.26 price per pound
$1.26 per lb. × 7 lbs. = $8.82 for cottage cheese

Sometimes, however, recipes created for only a small number of servings have the ingredients given in volume measures such as 3 tablespoons cottage cheese or 1 1/2 cups cottage cheese. This can pose a problem, especially when the prices are given by weight.

Let's say you need to find the cost of the following ingredients after adjusting them from 4 servings to 64 servings. Applying the formula, you find the conversion factor is 16. You proceed by multiplying.

Ingredient Amount	× 16	New Amount of Ingredient
3 tablespoons cinnamon	×16	48 tablespoons cinnamon
5 cups flour	×16	80 cups flour
2 cups sliced carrots	×16	32 cups sliced carrots
1 cup parmesan cheese	×16	16 cups parmesan cheese

How will you write the adjusted ingredients? You know that you don't want to write 48 tbsp. of cinnamon on your recipe card, so you proceed to convert.

The Good News Is: You learned how to convert measures in Chapter 12.

$$\frac{48 \text{ tbsp.}}{1} \times \frac{1 \text{ cup}}{16 \text{ tbsp.}} = 3 \text{ cups cinnamon}$$

The Bad News Is: You now need to convert from volume to weight.

OK, so you need 3 cups of cinnamon. What's the problem? The problem is that when you want to know the cost of cinnamon, you realize that the distributor gives the price of cinnamon in the **weight** measure of ounces. You have just converted the cinnamon ingredient using **volume measures,** as in **tablespoons** and **cups.** Now what?

Refer to the table *Food Weights/Volume Equivalents* found in the Appendix 1. This table gives some common food approximate weights and measures. Keep in mind that "a pint is a pound the world around," doesn't work for all foods!

Reading the table you find that 1½ cups of jelly weigh approximately 1 pound.

$$\text{Jelly} \Rightarrow 1 \text{ pound} \cong 1\tfrac{1}{2} \text{ cups}$$

EXAMPLE 5: Find the cost of three cups of ground cinnamon. The price given by the distributor is: *6 – 16 OZ CINNAMON $4.70*

Step 1: Referring to the table, you see that 4 tbsp of ground cinnamon weighs approximately 1 ounce.

Step 2: Convert, using the unity fraction method.

From the equivalents table.

$$\frac{3 \text{ cups cinnamon}}{1} \times \frac{16 \text{ tbsp}}{1 \text{ cup}} \times \frac{1 \text{ oz}}{4 \text{ tbsp cinnamon}} \times \frac{\$4.70}{6 \times 16 \text{ oz}}$$

$$\Rightarrow \quad (3 \times 16 \times \$4.70) \div (4 \times 6 \times 16)$$

$$\Rightarrow \quad 2.256 \div 96$$

$$\Rightarrow \quad \$0.5875$$

Step 3: The cost of 3 cups of cinnamon is approximately 59¢.

Find the cost of 32 cups of shredded carrots. Your distributor price is *20# FOR $12.90.*

Ingredient Amount	× 16	New Amount of Ingredient
3 tablespoons cinnamon	×16	48 tablespoons cinnamon
5 cups flour	×16	80 cups flour
2 cups shredded carrots	*×16*	*32 cups shredded carrots*
1 cup parmesan cheese	×16	16 cups parmesan cheese

Step 1: Referring to Appendix 1, you see that 3 cups of shredded carrots weigh approximately 1 pound.

Step 2: Convert, using the unity fraction method.

From the equivalents table.

$$\frac{32\ cups\ carrots}{1} \times \frac{1\ pound}{3\ cups} \times \frac{\$12.90}{20\ pounds}$$

$$\Rightarrow \quad 32 \times \$12.90) \div (3 \times 20)$$

$$\Rightarrow \quad 412.8 \div 60$$

$$\Rightarrow \quad \$6.88$$

Step 3: The cost of 3 cups of carrots is approximately $6.88.

EXAMPLE 7: Find the total cost, cost per pie, and the cost per portion, of 30 Layered Salmon Pies. You may assume the recipe is adjusted for the shrinkage of the salmon during cooking.

Ingredients for 30 Pies	Distributor Price	Your Cost Per Item
7.5 lb Salt pork	$6.15 per pound	?
30 lb Salmon steak	10# of 6 oz fillets $48.15	?
3¾ c Flour	$15.05 for 50#	?
7.5 c Celery	1 stalk $0.09	?
15 oz Onion	50 lb $12.40	?
30 Potatoes	50#, 100CT for $15.06	?
60 Pastry shells	20/9" for $14.65	?

Total Cost ___?___

Cost Per Pie ___?___

Cost Per Serving ___?___

Step 1: Find the cost per item.

a. $7.5 \times 6.15 = \$46.125 \cong \$46.13$ for salt pork
Multiply the total pounds of pork needed times the price per pound.

b. You need 3-10# boxes $\Rightarrow$ 3 x 48.15 = \$144.45 for salmon
Multiply the price per box times the 3 boxes.

c. Look at the weight/measure table. You find that 1 cup of bread flour weighs approximately 5 oz. Use this conversion fact as a unity fraction.

$$\frac{3\tfrac{3}{4}\ \text{cups flour}}{1} \times \frac{5\ \text{ounces}}{1\ \text{cup flour}} \times \frac{1\ \text{lb}}{16\ \text{oz}} \times \frac{\$15.05}{50\ \text{lb.}} \cong \$0.35\ \text{for flour}$$

d. You need 7½ cups of celery. According to the equivalency chart found in Appendix 1, you get approximately ¾ cup of chopped celery from one celery stalk. Use the unity fraction method.

$$\frac{7.5\ \text{cups of celery}}{1} \times \frac{1\ \text{stalk}}{3/4\ \text{cup of celery}} \times \frac{\$0.09}{1\ \text{stalk}}$$

$$\Rightarrow \frac{7.5 \times 0.09}{0.75} = 0.9 \cong \$0.90\ \text{for celery}$$

e. Find the price of onion per ounce. Use the unity fraction conversion method!

$$\frac{15\ \text{oz}}{1} \times \frac{1\ \text{lb}}{16\ \text{oz}} \times \frac{\$12.40}{50\ \text{lb}} \cong \$0.23\ \text{for onion}$$

f. Note that 100 potatoes cost \$15.06 $\Rightarrow$ 15.06 ÷ 100 = \$ 0.1506 per potato

$$\frac{30\ \text{potatoes}}{1} \times \frac{\$15.06}{100\ \text{potatoes}} \cong \$4.52\ \text{for potatoes}$$

g. You need 60 pie shells, 2 per pie (top and bottom). You note that 20 shells cost \$14.65.

$$\frac{60\ \text{shells}}{1} \times \frac{\$14.65}{20\ \text{shells}} \cong \$43.95\ \text{for shells}$$

Step 2: To find the total cost of the 30 pies, just add the individual costs for each ingredient.

pork   salmon   flour   celery   onion   potatoes   shells
46.13 + 144.45 + 0.35 + 0.90 + 0.23 + 4.52 + 43.95 = 240.53

$\Rightarrow$ The total cost of 30 pies is \$240.53

Step 3: To find the cost per pie remember that each recipe yields 30 pies.

$$\$240.53 \div 30 \text{ pies} = \$7.285 \cong \$8.02 \text{ per pie}$$

Divide the total cost for the pies by the number of pies to get the price per pie.

Step 4: To find the cost per portion remember that each pie yields 4 servings.

$$\$8.02 \div 4 \text{ pieces per pie} \cong \$2.01 \Rightarrow \$2.01 \text{ per serving}$$
$$2.005 \Rightarrow \$2.01$$

Divide the total cost per pie by the number of pieces to get the price per serving.

Ingredients for 30 Pies	Distributor Price	Your Cost Per Item
7.5 lb. Salt pork	$6.15 per pound	$ 46.13
30 lb. Salmon Steak	10# of 6oz fillets $48.15	$144.45
3 c Flour	$15.05 for 50#	$0.35
7.5 c Celery	1 stalk $0.09	$0.90
15 oz Onion	50 lb. $12.40	$0.23
30 Potatoes	50#, 100CT for $15.06	$4.52
60 Pastry Shells	20/9" for $14.65	$43.95

Total Cost	$240.53
Cost per Pie	$ 8.02
Cost per Serving	$ 2.01

Exercise Set 15

Be sure to put your answer in the "best" unit for the question. Do *not* leave answers such as *13.6 pounds*. Convert your answers into *13 pounds 10 ounces*. Keep in mind that you should always round up when purchasing ingredients for a recipe. Although 0.6 pounds converts to 9.6 ounces, round up to 10 oz. This practice will make sure you don't run short. If, after converting a recipe, you find that you need 16.23 bell peppers, round up to 17 peppers. Round 24.35 ounces to 25 ounces, and so on.

1. The portion yield for a chicken salad is 15 – 12 oz servings. What is the approximate bulk yield?

2. You were able to serve each of the 65 people 2 dinner rolls. What was the approximate bulk yield of the recipe you used for dinner rolls, if each roll weighed close to 1.4 ounces?

3. The bulk yield of a lamb hash recipe is 6 pounds. What would the portion yield be if each serving should contain 10 ounces?

4. The bulk yield of a Hungarian Goulash recipe is 108 pounds. Describe the portion yield if each person is to receive 12 ounces of goulash.

5. What size portion would each person receive, if you use the recipe from question 4 and are feeding 250 people?

6. Using the yield for hash from question 3, what size portion would each person receive if you were serving 20 people?

7. The bulk yield of a muffin recipe is 12 pounds 51/2 ounces. Describe the portion yield if each muffin is to have 3 ounces of batter.

8. After creating your own recipe for carrot soup, you find that the bulk yield is 4 gallons.

   a. How many people could you serve if your serving size was 1 cup?

   b. Approximately what size portion would you serve to each of 84 people?

9. Using ratio and proportion.
   a. You need 4 ounces of milk powder for a recipe that yields 8 dozen cookies. How much milk powder will you need for 14 dozen cookies?

   b. How much milk powder will you need to make 5 dozen cookies?

10. If you convert the soufflé recipe to yield 13 soufflés, how much flour would you need?

11. If you convert the soufflé recipe to yield 16 soufflés, how much salmon would you need?

12. How much butter would you need for 3 soufflés?

13. How much milk for 38 soufflés?

Salmon Soufflé
Bulk Yield: 8 soufflés

8 small cans of red salmon
1¼ doz. eggs, separated
7 oz butter
1/2 cup flour
8 large cups fresh breadcrumbs
2/3 cup parsley (chopped finely)
2 quarts of milk
salt and pepper to taste

---

14. You need 60 servings of your favorite coleslaw recipe for a picnic. Each serving should be approximately 3/4 cup. Your recipe lists the bulk yield as 1 gallon. How will you adjust the ingredient amounts? What will be your conversion factor?

15. Convert the following recipe. *C.F. stands for conversion factor.*

Ingredients for 90 Servings of Chili	Multiply by C.F.	Ingredients for 65 Servings of Chili
221/2 pounds of lean ground beef		
17 green bell peppers		
16 large onions, diced		
15 cloves garlic, minced		
18 oz chili powder		
15–141/2 oz cans tomatoes		
1/2 cup A1 Steak Sauce		
1 qt 3 cups water		
15–(151/2- to 16-ounce) cans kidney beans, drained		

16. Convert the following recipe for Ravioli.

Ingredients for 4 Servings of Ravioli	Multiply by C.F.	Ingredients for 25 Servings of Ravioli
1 (16-ounce) package cheese ravioli		
1 cup fat-free cottage cheese		
1/2 cup evaporated skim milk		
1 teaspoon dried rosemary		
1/4 teaspoon salt		
1/4 teaspoon ground black pepper		
2 teaspoons fresh lemon juice		
1/4 cup finely shredded fresh Parmesan cheese		
1 teaspoon finely shredded lemon peel		
Lemon wedges (optional)		
3 tablespoons snipped fresh chives		

Cheese Ravioli with Rosemary and Lemon

For questions 17–20, use the distributor price list found at the end of these exercises. Find the total cost, cost per serving and menu price, for one serving of each of the following recipes. To figure your menu price use a 31% food cost markup. *Refer to the weight and approximate equivalent in measure chart. You will need it when you are costing.*

17.

> ### *Fettuccini Alfredo*
>
> **Yield:** 36 servings
>
> 4 1/2 pounds fettuccine
> 48 oz evaporated skim milk
> 3 cups half-and-half
> 12 tablespoons butter, cut into small pieces
> 9 cups grated fresh parmesan cheese   *Hint: 1qt ~ 1 lb.*
> 12 tablespoons snipped fresh chives   *Hint: 1tbsp ~ 1/4 oz.*

Ingredients for	Distributor Price	Your Cost per Item

Total Cost _____

Cost per Serving _____

Menu Price _____

18.

BEEF TENDERLOINS WITH CREAM SAUCE

Makes 48 servings

48 (4-ounce) beef tenderloin steaks, 1 inch thick
6   ounces margarine
3   pounds dairy sour cream
3   cups A -1 Steak Sauce
2¼ cups milk
8   tablespoons all-purpose flour *(1 tbsp. ~ 1/4 oz)*
24 teaspoons snipped fresh thyme leaves or 1/2 teaspoon dried thyme, crushed

---

Ingredients for	Distributor Price	Your Cost per Item

Total Cost _____

Cost per Serving _____

Menu Price _____

19.

DIJON SHRIMP SCAMPI

Makes 48 servings

12   pounds large shrimp, cleaned and deveined
12   clove garlic, crushed
11½ cups butter
4   cups GREY POUPON COUNTRY DIJON Mustard
3   cups lemon juice
3   cups chopped parsley
Hot cooked rice, optional

Ingredients for	Distributor Price	Your Cost per Item

Total Cost _____

Cost per Serving _____

Menu Price _____

20.

Artichoke Lasagna

Makes 32 servings

8 lbs. lasagna noodles
3 tablespoons olive oil
4 medium onions, chopped
12 garlic cloves, minced
4 cups low-sodium vegetable stock
1 cup chopped fresh basil
4½ pounds frozen artichoke hearts,
partially thawed
½ teaspoon grated nutmeg
2 tablespoons margarine
2 tablespoons unbleached flour
6 cups evaporated skim milk

4 cups chopped roasted sweet red peppers
1 cup seasoned dry bread crumbs
6 tablespoons grated parmesan cheese
2½ pounds frozen chopped
spinach, partially thawed

Ingredients for	Distributor Price	Your Cost per Item

Total Cost _____

Cost per Serving _____

Menu Price _____

DISTRIBUTORS PRICE LIST

2 1/2 #	16-20 Round Gourmet Shrimp	$22.35
12-32oz	Lemon Juice Reconstituted	$15.12
24-5oz	A-1 Steak Sauce	$38.76
40-4oz	Beef Tenderloin	$82.95
6-1 oz	Chopped Chives	$ 3.50
4-5oz	Basil Leaf	$ 4.05
20#	Fettuccine Pasta	$14.05
10#	Lasagna Noodles	$ 7.65
6-7oz	Thyme Leaves	$ 5.60
20#	Chopped Onion	$12.40
6-1qt	Chopped Garlic	$ 4.40
50oz	Vegetable Stock	$ 2.60
5#	Gourmet Bread Crumbs	$ 2.05
15doz	Eggs	$ 8.20
48-12oz	Evaporated Skim Milk	$29.95
25#	Flour	$ 6.45
12-9oz	Grey Poupon Mustard	$ 9.85
4gal	Real Lemon Juice	$33.25
12-1#	Parmesan Cheese	$ 3.70
12oz	Chopped Red Peppers	$ 5.00
6-16oz	Nutmeg	$ 7.60
30#	Margarine	$13.75
36#	Butter	$53.75
20#	Olive Oil	$35.00
3#	Chopped Spinach	$ 1.90
3-5#	Sour Cream	$ 4.55
1gal	Half & Half Cream	$10.62
6-1gal	Whole Milk	$20.38
4/2.5#	Artichoke Hearts	$75.20
11 oz	Parsley Flakes	$ 7.95

CHAPTER 16

Bakers' Formulas

Let's go back to Charlie's, they have great bread with honey butter,   hmmmm...

The process of baking rolls or individual loaves of bread day after day, and having them consistently taste great, look good, and feel majestic, is what makes some restaurants famous. Many people frequent a particular restaurant for these "extras."

But how do you make sure the loaves are consistent when you have to constantly change the amount of loaves to bake? On Sundays you need 650 loaves, while on Monday you may need only 150 loaves. Can one recipe be scaled to fit all needs?

BAKER'S PERCENTAGE

The *baker's percentage* is used to convert the weights of the ingredients in a given baking formula into a percent of the total weight of the flour in the given formula.

Whoa! Better read that again . . .

Now let's see what an expert has to say about it! Here is what Rosada Didier, of the National Baking Center, says about baker's percentage.

Why is the baker's percentage important for the baker?
- Consistency in production
- Easy calculation absorption rate of the flour
- Simple increase or decrease in dough size using the same formula
- Ease in comparing formulas
- Ability to check if a formula is well-balanced
- Ability to correct defects in the formula

Important characteristics of the baker's percentage
- The baker's percentage is always based on the total weight of the flour in the formula.
- Flour is always represented by the value of 100%; i.e., all other ingredients are calculated in relationship to the flour.
- The baker's percentage can only be calculated if the amount of all the ingredients in the formula are expressed in the same unit of measure; for example, you cannot mix grams and ounces, or pounds and kilograms, in the same formula.
- Units of measure must be expressed in terms of weight, not volume; for example, you cannot mix pounds and quarts in the same formula. (Exception: metric system)
- The baker's percentage works best with the metric system, because metrics are based on units of 10, as are percentages (e.g., 100 = 10 x 10).

The flour portion in a baker's formula is always represented by the value 100%. All other ingredients are calculated in relationship to the flour.

Let's look at a basic dough formula.

Flour	50 kg
Water	30 kg
Salt	1 kg
Yeast	.75 kg

For this formula, 50 kg of flour represents 100%. Think back to our discussion in Chapter 5 about percentage and the *part to the whole* concept.

If the "whole" is represented by the flour, then the water is what part of the whole? 30:50, or 30 parts water for every 50 parts flour.

We write this ratio as $\frac{30}{50}$ or $30 \div 50 = 0.6$ or 60 %

So in this formula the baker's percent for the water is 60%. This means the weight of the water represents 60% of the weight of the flour.

Example 1: Convert the following dough formula from weight to baker's percent.

Flour : 34 kg
Water : 23 kg
Salt : 680 g
Yeast : 340 g

Step 1: Before doing any calculations, you must convert the ingredients to a common unit of measure. In this formula there are kilograms and grams. Convert the grams to kilograms.

Flour : 34 kg
Water : 23 kg
Salt : 0.68 kg
Yeast : 0.34 kg

Step 2: Next express each ingredient as a *part to the whole* ratio, using 34 kg of flour as the whole, and then divide.

Flour	34 : 34	$\Rightarrow$	34 ÷ 34	=	1
Water	23 : 34	$\Rightarrow$	23 ÷ 34	$\cong$	0.676
Salt	0.68 : 34	$\Rightarrow$	0.68 ÷ 34	=	0.02
Yeast	0.34 : 34	$\Rightarrow$	0.34 ÷ 34	=	0.01

Step 3: Convert the decimals to percents, then write the complete formula with baker's percents.

Flour : 34 kg 100%
Water : 23 kg 67.6%
Salt : 680 g 2 %
Yeast : 340 g 1%

This is the complete formula with baker's percents.

Example 2: Use the baker's percents found in Example 1 to make a dough that requires 45 kilograms of flour.

Step 1: We know that for this formula 45 kg of flour represents 100%. Calculate the amount needed for each ingredient based on the baker's percents.

Flour	100% of	45 kg	$\Rightarrow$	1 x 45	=	45 kg
Water	67.6% of	45kg	$\Rightarrow$	0.676 x 45	=	30.42 kg
Salt	2% of	45kg	$\Rightarrow$	0.02 x 45	=	0.9 kg (900 g)
Yeast	1% of	45kg	$\Rightarrow$	0.01 x 45	=	0.45 kg (450 g)

Step 2: Write the complete formula with baker's percents.

Flour : 45 kg 100%
Water : 30.42 kg 67.6%
Salt : 900 g 2 %
Yeast : 450 g 1%

Example 3: You are asked to fill a production order:

50 baguettes	@	350 g of dough
40 balls	@	400 g of dough
300 rolls	@	80 g of dough

All of these breads will be made from the same dough. The baker's percentages for the formula are:

Flour	:	100%
Water	:	67%
Salt	:	2%
Yeast	:	1%

Step 1: Determine the total amount of dough needed.

Baguettes	50 x 350 g	=	17,500 g	=	17.5 kg
Balls	40 x 400 g	=	16,000 g	=	16 kg
Rolls	300 x 80 g	=	24,000 g	=	24 kg
Total Dough Needed		=	57,500 g	=	57.5 kg

Step 2: Determine the amount of flour needed to make 57.5 kg of dough. First find the total baker's percentages needed for this formula by adding.

Flour	:	100 %
Water	:	67 %
Salt	:	2 %
Yeast	:	1 %
Total		170 %

Step 3: What part of the total dough is represented by flour? *Part to the whole.*

$$\frac{flour}{total\ dough} \quad \frac{100\%}{170\%} = 0.5882 \quad or \quad 58.82\%$$

Step 4: Calculate the amount of flour needed. From Step 3 above we know:

58.82% of the total dough needed is flour

or

.5882 x 57.5 kg = 33.82 kg

Step 5:   Determine the desired weights of the remaining ingredients.

Note: To simplify the calculations, and to make sure we produce enough dough, we always round the amount of flour up to the next whole number.

Flour	100%	34 kg	*(rounding up from 33.82)*
Water	67%	0.67 × 34 kg	= 22.78 kg
Salt	2%	0.02 × 34 kg	= 0.68 kg or 680 g
Yeast	1%	0.01 × 34 kg	= 0.34 kg or 340 g

Step 6:   Verify your calculation by totaling the weights of the ingredients.

Flour	:	34.00 kg	100%
Water	:	22.78 kg	67%
Salt	:	00.68 kg	2%
Yeast	:	00.34 kg	1%
Total		57.80 kg	

Although only 57.5 kg of dough are needed, we will have 57.80 kg with a small amount of extra dough, only 300 g, due to rounding the amount of flour to the next whole number.

Exercise Set 16

Remember:	1,000 g = 1 kg	·	1 kg water = 1 liter	·	1 pint water = 1 pound

1. The flour portion in a baker's formula is always represented by _____%.

2. The baker's percentage is always based on the total weight of the _____ in the formula.

3. A certain formula requires 56 kg of flour and 37 kg of water. What is the baker's percent for the water?

4. Solana is using a formula that requires 15 kg of flour and 0.3 kg of salt. For this formula, the weight of the salt represents what percent of the weight of the flour?

For questions 5–8: Convert the following dough formulas from weight to baker's percent, then write the complete formula with baker's percents. *Remember to express the ingredients in the same **unit** of measure before you calculate.*

5. Flour : 21 kg _____
   Water : 13.23 kg _____
   Salt   : 315 g _____
   Yeast : 210 g _____

6. Flour : 65 lb. _____
   Water : 42.25 lb. _____
   Salt   : 3 lb _____
   Yeast : 1 lb _____

7. Flour : 30 lb. _____
   Water : 18 lb. _____
   Salt   : 9 oz. _____
   Yeast : 5 oz. _____

8. Flour : 18 kg _____
   Water : 12.24 kg _____
   Salt   : 450 g _____
   Yeast : 184 g _____

For questions 9–12: Using the following baker's percents, write a *complete* formula to make a dough that requires the given amount of flour.

Flour   : 100%
Water   : 72 %
Salt   : 2.5 %
Culture : 50 % (leaven for sourdough)

9.   Flour : 40 kg   11.   Flour : 12 kg

10.   Flour : 63 kg   12.   Flour : 23 lb.

You are asked to fill production orders given the following information.
   a. Determine the necessary amount of flour.
   b. Find the desired weights of the remaining ingredients.
   c. Verify your calculations to determine your accuracy.

All of these breads will be made from the same dough. The baker's percentages for the formula are:

Flour	:	100%
Water	:	70%
Salt	:	2%
Yeast	:	1%

Production Order #13 :

15 baguettes	@	340 g of dough
150 rolls	@	75 g of dough
5 balls	@	375 g of dough

Production Order #14:

19 baguettes	@	370 g of dough
250 rolls	@	85 g of dough
12 balls	@	425 g of dough

APPENDIX 1

Food Weight/Volume Equivalents

Food/Measure	Weight	Volume
All-purpose flour	1 pound	3 ½ - 4 cups
Almonds in shell	1 pound	1 to 1 ¼ cups
Almonds, shelled	1 pound	3 cups nutmeats
Apples	1 pound	3 medium (3 cups sliced)
Apples, 3 medium	1 pound	3 cups sliced
Bananas	1 pound	3 medium (1 ½ cups mashed)
Basil	1tsp dry	3 tsp fresh
Beans, kidney, 1 cup dry	½ pound	2-½ cups cooked
Beans, lima, 1-¼ cup dry	½ pound	3 cups cooked
Beans, navy, 1 cup dry	½ pound	2-½ cups cooked
Beets	1 pound	2 cups sliced
Brazil nuts in shell	1 pound	1 ½ cups nutmeats
Brazil nuts, shelled	1 pound	3 ¼ cups nutmeats
Bread crumbs	4 ounces	1 cup
Bread	1 pound	20 to 22 slices
Broccoli	1 pound head	2 cups flowerets
Brown sugar	1 pound	2 ¼ cups, packed
Butter or margarine	¼ pound	½ cup
Cabbage	1 pound head	4 ½ cups shredded
Carrots	1 ounce	4 tablespoons
Carrots, 1 pound	3 cups shredded	2-½ cups diced
Cauliflower	1-½ pound head	2 cups cooked
Celery	1 large stalk	¾ cup sliced
Celery	1 large stalk	¾ cup diced
Cheese	hard 3 ounces	1 cup shredded
Cheese	soft 4 ounces	1 cup shredded
Chives	1 teaspoon dry	3 teaspoons fresh
Chocolate	chips 1 cup	6 ounces
Chocolate	baking 1 square	1 ounce
Cinnamon	1 oz	4 tablespoons
Coffee	1 pound	80 tablespoons
Confectioner's sugar	1 pound	about 3 ¾ cups
Corn	2 medium ears	1 cup kernels
Corn	10 ounces	2 cups
Corn syrup	16 ounces	2 cups
Cornmeal	1 pound	3 cups

Food/Measure	Weight	Volume
Cottage cheese	8 ounces	1 cup
Egg whites	1 cup	8-10 whites
Egg yolks	1 cup	12 to 14 yolks
Evaporated milk	5-ounce can	⅔ cup
Evaporated milk	12-ounce can	1 ⅔ cup
Flaked coconut	3 ½-ounce can	1 ⅓ cups
Flour	5 ounces	1 cup
Fresh breadcrumbs	1 slice bread with crust	½ cup
Frozen vegetables	1 pound	3 cups
Garlic	3 large cloves	1 tablespoon minced
Gelatin	1 envelope	1 tablespoon
Graham crackers	14 squares	1 cup crumbs
Granulated sugar	1 pound	2 ¼ to 2 ½ cups
Green beans	fresh 1 pound	2-½ cups cooked
Green pepper	1 large	1 cup diced
Hazelnuts in shell	1 pound	1 ½ cups nutmeats
Hazelnuts, shelled	1 pound	3 ½ cups nutmeats
Heavy or whipping cream	1 cup	2 cups whipped
Herbs	1 teaspoon dry	3 teaspoons fresh
Honey	16 ounces	1 ⅓ cups
Jelly	1 pound	1 ½ cups
Lemon	1 medium	3 tablespoons lemon juice & 1 tablespoon grated lemon peel
Lemon	1 whole	3 tablespoons juice
Lettuce	1 pound head	6 ¼ cups torn
Lime	1 medium	2 tablespoons lime juice & 1 teaspoon grated lime peel
Lime	1 whole	2 tablespoons juice
Maple syrup	12 ounces	1 ½ cups
Mushrooms	3 cups raw	1 cup cooked
Mushrooms	fresh 1 pound	4 cups chopped
Mustard-ground	3 ¼ ounce	1 cup
Mustard-prepared	4 oz	1 cup
Nutmeg	¼ oz	1 tablespoon
Oats, quick cooking	1 cup	1 ¾ cups cooked
Olive Oil	1 pound	2 cups
Onion	1 medium	½ cup chopped
Onions	1 large	1 cup chopped
Onions	1 pound	3 large
Orange	1 medium	⅓ to ½ cup juice and 2 tablespoons grated peel
Parmesan cheese	1 ounce	4 tablespoons
Peach	2 medium	1 cup sliced
Peanuts	1 pound shelled	4 cups
Pear	2 medium	1 cup sliced
Pears, 1 medium	4 ounces	½ cup sliced
Pecans	3-½ ounces halves	1 cup
Pecans in shell	1 pound	2 ¼ cups nutmeats
Pecans, shelled	1 pound	4 cups nutmeats
Peppers, bell, 1 large	6 ounces	1 cup diced
Peppers, red	6 ounces	1 cup diced
Potatoes	3 medium	1 pound
Raisins	1 pound	3 cups, loosely packed
Rice	brown 1 cup raw	3-⅛ cups cooked

(continued)

Food/Measure	Weight	Volume
Rice	white 1 cup raw	3 cups cooked
Rice	wild 1 cup raw	4 cups cooked
Rice, long grain, precooked	1 cup	2 cups cooked
Saltine crackers	28 carackers	1 cup crumbs
Semisweet chocolate	6 ounces	1 cup
Shredded coconut	7 ounce bag	2 ⅔ cups
Sour cream	8 ounces	1 cup
Spinach, 1 pound, fresh	6 cups leaves	1-¾ cups cooked
Squash, winter	1 pound	1 cup mashed
Strawberries	1 quart	4 cups sliced
Sweet potatoes	3 medium	3 cups sliced
Sweetened condensed milk	14 ounce can	1 ¼ cups
Thyme	8 tbsp	4 ounces
Tomato	2 ½ pounds	3 cups seeded, chopped, and drained
Tomato paste/sauce	8 ounces	1 cup
Unsweetened chocolate	1 ounce	1 square
Vanilla wafers	22 wafers	1 cup crumbs
Walnuts in shell	1 pound	2 cups nutmeats
Walnuts, halves	3-½ ounces	1 cup
Walnuts, shelled	1 pound	4 cups nutmeats
White potatoes	3 medium	2 cups cubed or 1 ¾ cups mashed
Yeast, 1 envelope	1 tablespoon	¼ ounce
Zucchini, 1 pound	3 cups sliced	2-½ cups chopped

VOLUME TO WEIGHT

ONE CUP OF	ENGLISH	METRIC
1 egg, AA large	2 ounces	57 grams
All-purpose flour	4-¼ ounces	121 grams
Almonds, chopped	3 ounces	85 grams
Almonds, ground	3-¾ ounces	107 grams
Almonds, sliced	3 ounces	85 grams
Almonds, slivered	4-¼ ounces	119 grams
Almonds, whole	6 ounces	168 grams
Bread flour	4-½ ounces	130 grams
Butter, unsalted (2 sticks)	8 ounces	227 grams
Cake flour	4 ounces	114 grams
Cocoa butter	9 ounces	256 grams
Corn starch	4-¼ ounces	120 grams
Corn Syrup	11-½ ounces	328 grams
Cream, heavy or whipping	8 ounces	232 grams
Dutch-processed cocoa	3-¼ ounces	92 grams
Hazelnuts, whole	5 ounces	140 grams
Honey	12 ounces	340 grams
Milk, buttermilk	8-½ ounces	242 grams
Nonalkalized cocoa	3 ounces	89 grams
Sour cream, half-&-half	8-½ ounces	242 grams
Sugar, dark brown, packed	8-½ ounces	240 grams
Sugar, granulated or superfine	7 ounces	200 grams
Sugar, light brown, packed	7-⅔ ounces	213 grams
Sugar, powdered	4 ounces	115 grams
Vegetable shortening	6-¾ ounces	191 grams
Walnuts & pecans, chopped	4 ounces	115 grams
Walnuts & pecans, halves	3-½ ounces	100 grams
Water	8-⅓ ounces	236 grams
Whole wheat flour	4-⅔ ounces	140 grams

APPENDIX 2

Conversion Chart

METRIC CONVERSIONS

To convert this	To this	Multiply by
Length		
inches	millimeters (mm)	25.4
feet	centimeters (cm)	39
yards	meters (m)	00.91
miles	kilometers (km)	01.61
millimeters	inches	00.04
centimeters	inches	00.4
meters	inches	39.37
meters	yards	01.1
kilometers	miles	00.6
Weight		
ounces	grams (g)	28
pounds	kilograms (kg)	00.45
short tons	metric tons	00.9
kilograms	pounds	02.2
metric tons	pounds	2,204.6
metric tons	short tons	01.1
Area		
square inches	square centimeters	06.5
square	square meters	00.09
square miles	square kilometers	02.6
acres	hectares	00.4
square centimeters	square inches	00.16
square meters	square yards	01.2
square kilometers	square miles	00.4
hectares	acres	02.5

To convert this	To this	Multiply by
Volume		
teaspoons	milliliters	5
tablespoons	milliliters	15
fluid ounces	milliliters	30
cups	liters	00.24
pints	liters	00.47
quarts	liters	00.95
gallons	liters	03.8
cubic feet	cubic meters	00.03
cubic yards	cubic meters	00.76
milliliters	fluid ounces	00.03
liters	pints	02.1
liters	quarts	01.06
liters	gallons	00.26
cubic meters	cubic feet	35
cubic meters	cubic yards	01.3
Temperature		
Fahrenheit	Celsius	.56 (after subtracting 31)
Celsius	Fahrenheit	1.82 (then add 32)
Farm Products		
pounds per acre	kilograms per hectare	1.14
short tons per acre	kilograms per hectare	2.25
kilograms per hectare	metric tons per hectare	.001
kilograms per hectare	pounds per acre	.88
tons per hectare	short tons per acre	.44
tons per hectare	kilograms per acre	1,000

BUSHEL/WEIGHT CONVERSIONS

1 bushel of	Weight in pounds	Weight in kilograms
wheat, soybeans, potatoes	60	27
corn, grain sorghum, rye, flaxseed	56	25
beets, carrots	50	23
barley, buckwheat, peaches	48	22
oats, cottonseed	32	14

1 metric ton of	Weight in pounds	Number of bushels
wheat, soybeans, potatoes	2,204.6	36.74
corn, grain sorghum, rye, flaxseed	2,204.6	39.37
beets, carrots	2,204.6	44.09
barley, buckwheat, peaches	2,204.6	45.93
oats, cottonseed	2,204.6	68.89

APPENDIX 3

Answers to Odd-Numbered Exercises

CHAPTERS 1 & 2

1. 6/7
3. 1 ¼
5. 1/3
7. 3 ½
9. 1/4
11. 8 ⅚
13. 2 ⅜
15. 7/20
17. 2/21
19. 5/7
21. 2 ½
23. 3
25. 1
27. 1 ½
29. 60
31. 2 ¼
33. 5.7333
35. 5.531
37. 1.085
39. 0.8505
41. 0.2252
43. 0.1
45. 0.01
47. 0.0001
49. 0.005
51. 0.$\overline{83}$
53. 1.571428
55. 4 2/11, 46/11, 4.$\overline{18}$
57. 6/7

CHAPTER 3

Answers are rounded to two decimal places.

1. a) No b) Yes c) No d) Yes e) Yes f) Yes
3. a) x = 30
   b) x = 1
   c) x = 399.75
   d) x = 2040
   e) x = 0.85
   f) x ~ 11.19
   g) x = 3.25
   h) x = 63
   i) x = 659.02
   j) x = 208.33
   k) x = 1
   l) x = 6

CHAPTER 4

1. 0.03
3. 3.45
5. .046
7. .0045
9. 1
11. 0.00006
13. 2.345
15. 160%
17. 0.004
19. 0.359

CHAPTER 4 (*cont.*)

21. $11,310.00
23. $10.71
25. $8.29
27. $360.00
29. $11,520 labor
    $21,600 Food Costs
31. $504
33. 74.29°
35. 28.7%
37. 99 gallons
39. $48,000
41. 24.74%
43. 16.54%

CHAPTER 5

1. 75%
3. 15%
5. 60$\overline{0}$%
7. 16.6%
9. 50%
11. 8$\overline{3}$%
13. 1.3%
15. 33 1/3 %
17. $660.00
19. a. 46.43% b. 53.57%
21. 33.3%
23. Total weight 172 pounds
    1. 16.28%
    2. 23.26%
    3. 11.63%
    4. 27.33%
    5. 15.12%
    6. 6.4%
    7.

CHAPTER 6

1. $240
3. $735
5. $56
7. $40.48
9. $103.95
11. $630
13. $2,763.91

15. a. $8,764.11
    b. $1,204.11
17. $3935.93; $535.93
19. $220
21. $0.80
23. a. 1.541% b. 0.05067%
25. $606.01

CHAPTER 7

1. To compare parts of a whole quantity, using %.
3. Bar graph.
5. Answers will vary.
7. Answers will vary.
9. a. horizontal, years, vertical, spending.
   b. student spending in the cafeteria.
   c. Yes, each wedge would represent a ratio of student spending for one year : to the total student spending for the four years. The whole pie would represent four years of student spending in the cafeteria.
11. a.

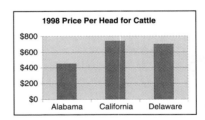

11. b.

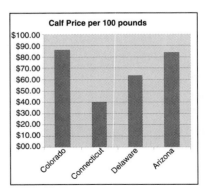

11 c.

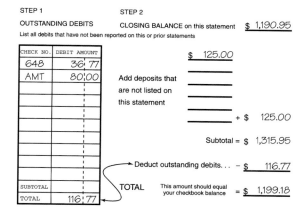

Total Value of Cattle in 1998

13. Show pie graph

Ayrshire	2	2.22%
Brown Swiss	2	2.22%
Guernsey	1	1.11%
Holstein	79	87.78%
Jersey	1	1.11%
Red/White	5	5.56%
Total	90	100.00%

15.

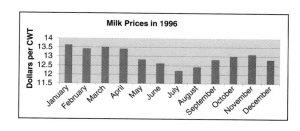

Milk Prices in 1996

17. 2500 CWT or 250,000 pounds
19. 15,000 CWT or 1,500,000 pounds
21. 5,000 lbs, 29,000 lbs.
23. insert pie graph
25. North Dakota
27. CA,MT,CO,UT,SD
29. Texas, Wyoming
31. a. 490,000 b. Breeding, 78.57% Market Bound 21.43% c. insert pie graph
33. 1997 and 1998
35. 3.3 million tons
37. insert pie graph

CHAPTER 8

1. a. eleven b. forty c. ninety d. eighty-eight e. fourteen f. forty-nine
3. Net Deposit: $931.65
5. Net Deposit: $356.35
13. Last line balance: $1,006.96
15. Last line balance: $5,386.03
17. Register Balance: $1,199.18

STEP 1	STEP 2	
OUTSTANDING DEBITS	CLOSING BALANCE on this statement	$ 1,190.95

List all debits that have not been reported on this or prior statements

CHECK NO.	DEBIT AMOUNT
648	36 77
AMT	80 00
SUBTOTAL	
TOTAL	116 77

$ 125.00

Add deposits that are not listed on this statement

+ $ 125.00

Subtotal = $ 1,315.95

Deduct outstanding debits... − $ 116.77

TOTAL This amount should equal your checkbook balance = $ 1,199.18

CHAPTER 9

1. False
3. True
5. False
7. True
9. True
11. g
13. e
15. f
17. d
19. 480 oz, $0.05
21. 480 oz, $0.04
23. 384 oz, $0.04
25. $0.06
27. $20.32
29. $0.04
31. see top next page

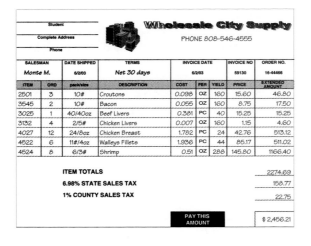

Student								
Complete Address					**Wholesale City Supply**			
Phone					PHONE 808-546-4555			

SALESMAN	DATE SHIPPED		TERMS		INVOICE DATE		INVOICE NO	ORDER NO.
Monte M.	6/2/03		Net 30 days		6/2/03		59130	16-44466

ITEM	ORD	pack/size	DESCRIPTION	COST	PER	YIELD	PRICE	EXTENDED AMOUNT
2501	3	10#	Croutons	0.098	OZ	160	15.60	46.80
3545	2	10#	Bacon	0.055	OZ	160	8.75	17.50
3025	1	40/40oz	Beef Livers	0.381	PC	40	15.25	15.25
3132	4	2/5#	Chicken Livers	0.007	OZ	160	1.15	4.60
4027	12	24/8oz	Chicken Breast	1.782	PC	24	42.76	513.12
4522	6	11#/4oz	Walleye Fillets	1.936	PC	44	85.17	511.02
4524	8	6/3#	Shrimp	0.51	OZ	288	145.80	1166.40

ITEM TOTALS	2274.69
6.98% STATE SALES TAX	158.77
1% COUNTY SALES TAX	22.75
PAY THIS AMOUNT	$ 2,456.21

CHAPTER 10

1. a. Subtotal $46.20, tax $3.23, pay amount $49.43 b. $7.41 c. $9.88 d. $50.57 e. $1.14
3. a. Subtotal $77.95, tax $5.46, pay amount $83.41 b. $12.51 c. $16.68 d. $16.59 e. $1.92
5. Subtotal $49.75 ,tax $3.48, pay amount $53.23 b. $7.98 c. $10.65 d. $46.77 e. $1.22
7. $1.46
9. $3.11, $9.33, $12.44
11. a. $248.71 b. $32.70 c. 80.00 d. $9.40 e. $1.75 f. 7.40
13. a. 3.86 b. 750 c. $126.22 d. 0.98 e. 90 f. $7.10
15. a. 600 b. 5 c. 40 d. 53.7 e. 3 f. 60 g. 15 h. 27 i. 0.3 j. 8 k. 4.58 l. 900 m. 5,000 n. 140

CHAPTER 11

1. 10.6%
3. 7.26%
5. 6.8%
7. 56.76%
9. 4.60%

11.

Arties Restaurant Corporation		23802
EMPLOYEE NANE: VALERIE STATHAM		EMPLOYEE ID: Hula Girl
SOCIAL SEC. #: 505-67-1234 RATE REG:	9.00	CHECK NUMBER: 23888
PAY PERIOD: 12/16/01 - 12/30/01		CHECK DATE: 1/07/02
REG: HOURS: 68.5 PAY	616.50	Tips Deducted - Lunch 850.00 -
O / T: HOURS 0.00 PAY		OTHER DEDUCTION 2 0.00
Tips Reported-Lunch	850.00	OTHER DEDUCTION 3 0.00
		OTHER DEDUCTION 4 0.00
--- GROSS PAY ----	1466.50	
FEDERAL INCOME TAX:	146.65 -	
FICA:	108.08 -	------ TOTAL WITHOLDING --- 1207.53 -
STATE INCOME TAX:	102.80 -	
LOCAL INCOME TAX:	-	----------- NET PAY ------ 258.97

13. 16.47%
15. 6.35%
17. 10.80%
19. 5.97%, 8.33%
21. No they were not. Property tax reads 0 for December but $250 for the year.
23. 0.846%
25. Answers will vary.

CHAPTER 12

1. 16 oz
3. 32 oz
5. 16 cups
7. 48 tsp
9. 2 Tbsp
11. 12 qt
13. 10 cups
15. 64 oz
17. 16 Tbsp
19. 15 cups
21. 13 cups
23. 1000 g
25. 6000 ml
27. 8.282 oz
29. 360 ml
31. 12.18 oz
33. 480 ml
35. 4.26 l
37. .30.8 lbs.
39. 2.27 kg
41. 2 cups
43. a. 20pt b. 10qt c. 2.5 gal d. 9.46 *l*

45. 3,913 pounds
47. 2,000 lbs or 1 T, 940 liters
49. 0.705 liters
51. 1.47 Tbsp (1Tbsp plus1_ tsp)
53. 1.93 pounds or 1 lb 14.84 oz

CHAPTER 13

1. 4lb 14 oz
3. 10 lb 10 oz
5. 23 lb 13 oz
7. 2 lb 15 oz
9. 1 lb 10 oz
11. 3 lb 3 oz
13. 6 gal 1 pt 1 c
15. 7g 1 qt 1 c
17. 2 gal 2 qt 1 pt 1 c
19. 13.695 l
21. 692.527 kg
23. 2 lbs 12 oz
25. 2 qt 1 pt 1 1/8 cup
27. 2 lb 12 oz
29. 6.4 oz
31. 437 oz
33. Yes, 20 oz
35. 2 pounds

CHAPTER 14

lettuce: 1) 1.175 2) 1.57 3) 0.79
almonds: 4) 5.726 5) 5.73 6) 0.72
Ahi steak: 7) 1.377 8) 1.72 9) 3.44
ribeye: 10) 7.985 11) 8.775 12) 3.29

13. $ 0.15
15. $ 0.30
17. $3.34
19. 216 left
21. $1.09
23. $4.18
25. $0.41
27. $0.02
29. 38%, 2.63
31. 41%, 2.44
33. 29.5%, 3.39
35. $1.04

37. $1.00
39. 4.86
41. a. 0.98 b. $1.44 c. $3.80 d. $6.06 e. $3.16 f. $4.65
43. 6/#10 can
45. $0.35
47. $0.56
49. $0.94
51. 3,195 ml
53. 473 ml

CHAPTER 15

1. 11 lb 4 oz
3. 9 portions
5. ≅ 7 oz per person
7. 65 portions
9. a. 7oz b. 2.5 oz
11. 16 cans salmon
    2 ½ doz eggs
    14 oz butter
    1 cup flour
    16 cups bread crumbs
    1 ⅓ cups parsley
    4 quarts milk
13. 2 gal 1½ qt
15.

Ingredients for 90 Servings of Chili	Ingredients for 65 Servings of Chili
22½ pounds of lean ground beef	16 ¼ pounds
17 green bell peppers	12 ¼ peppers
16 large onion, diced	12 onions
15 clove garlic, minced	11 cloves
18 oz chili powder	13 oz chili powder
15 - 14 ½ oz cans tomatoes	11 cans
½ cup A1 Steak Sauce	3 oz
1 qt 3½ cups water	5 ½ cups
15 - (151/2 to 16-ounce) cans kidney beans, drained	11 cans

17. Rounded off only at last step of calculations.

Ingredients for Fettuccini	Distributor Price	Your Cost per Item
4 ½ lb fettuccine	$14.05/20 lbs	$3.16
48 oz evaporated milk	$29.95/576 1 can = 1 ⅔ cup	$2.50
3 cups half and half	$10.62/gal	$1.99
12 tbsp butter (3/4 c)	$53.75/36 lbs 1/4 lb = 1/2 cup	$0.56
9 cups grated Parmesan	$3.70/12 lbs 4 oz = 1 cup	$0.69
12 tbsp chives	$3.50/6 oz	$1.75
	Total Cost	$10.65
	Cost per Serving	$ 0.30
	Menu Price	$ 0.95

19. Rounded off only at last step of calculations.

Ingredients for Shrimp Scampi	Distributor Price	Your Cost per Item
12 lb shrimp	$22.35/2 ½ lb	$107.28
12 clove garlic (1/4 c)	$4.40/6 qt	$0.05
1 ½ c butter	$53.75/36 lb 1/4 lb = ½ cup	$1.12
4 c mustard (16 oz)	$9.85/108 oz	$1.46
3 c lemon juice	$33.25/4 gal	$1.56
3 c parsley (9 oz)	$7.95/11 oz	$6.50
	Total Cost	$117.97
	Cost per Serving	$ 2.46
	Menu Price	$ 7.90

CHAPTER 16

1. 100%
3. 66.07%
5. 100%, 63%, 1.5%, 1%
7. 100%, 60%, 1.87%, 1.04%
9. flour: 40 kg : 100%
   water: 28.8 kg : 72%
   salt: 1 kg : 2.5%
   culture: 20 kg : 50%
11. flour: 12 kg : 100%
   water: 8.64 kg : 72%
   salt: 0.3 kg : 2.5%
   culture: 6 kg : 50%
13. dough needed 18.225 kg
   a. 11 kg flour
   b. water 7.7 kg, salt 220 g, yeast 110 g
   c. 11 kg +7.7 g + 0.220 g + 0.110 g = 19.03 kg

Index

Simple interest, 48
Solving equations, 16

T

Time, 40
Tipping guidelines, 106
Tips
    and gratuities, 108
    guidelines for tipping, 106
    pooling, 107
Total revenue, 123

U

U.S. Customary Measurement System, 132
U.S. orange production, bar graph, 69
Unity fractions, conversion using, 133

V

Variable, 12

W

Wages, 119
Weights and measures
    adding, 142–146
    converting, 131–141
Wool production, top producing states (1998),
        pie graph, 67
Writing a check, 74

Y

Yield, 147
    bulk, 163
    portion, 163
    recipe, 163